Ichetucknee

UNIVERSITY PRESS OF FLORIDA

Florida A&M University, Tallahassee
Florida Atlantic University, Boca Raton
Florida Gulf Coast University, Ft. Myers
Florida International University, Miami
Florida State University, Tallahassee
New College of Florida, Sarasota
University of Central Florida, Orlando
University of Florida, Gainesville
University of North Florida, Jacksonville
University of South Florida, Tampa
University of West Florida, Pensacola

University Press of Florida

Gainesville

Tallahassee

Tampa

Boca Raton

Pensacola

Orlando

Miami

Jacksonville

Ft. Myers

Sarasota

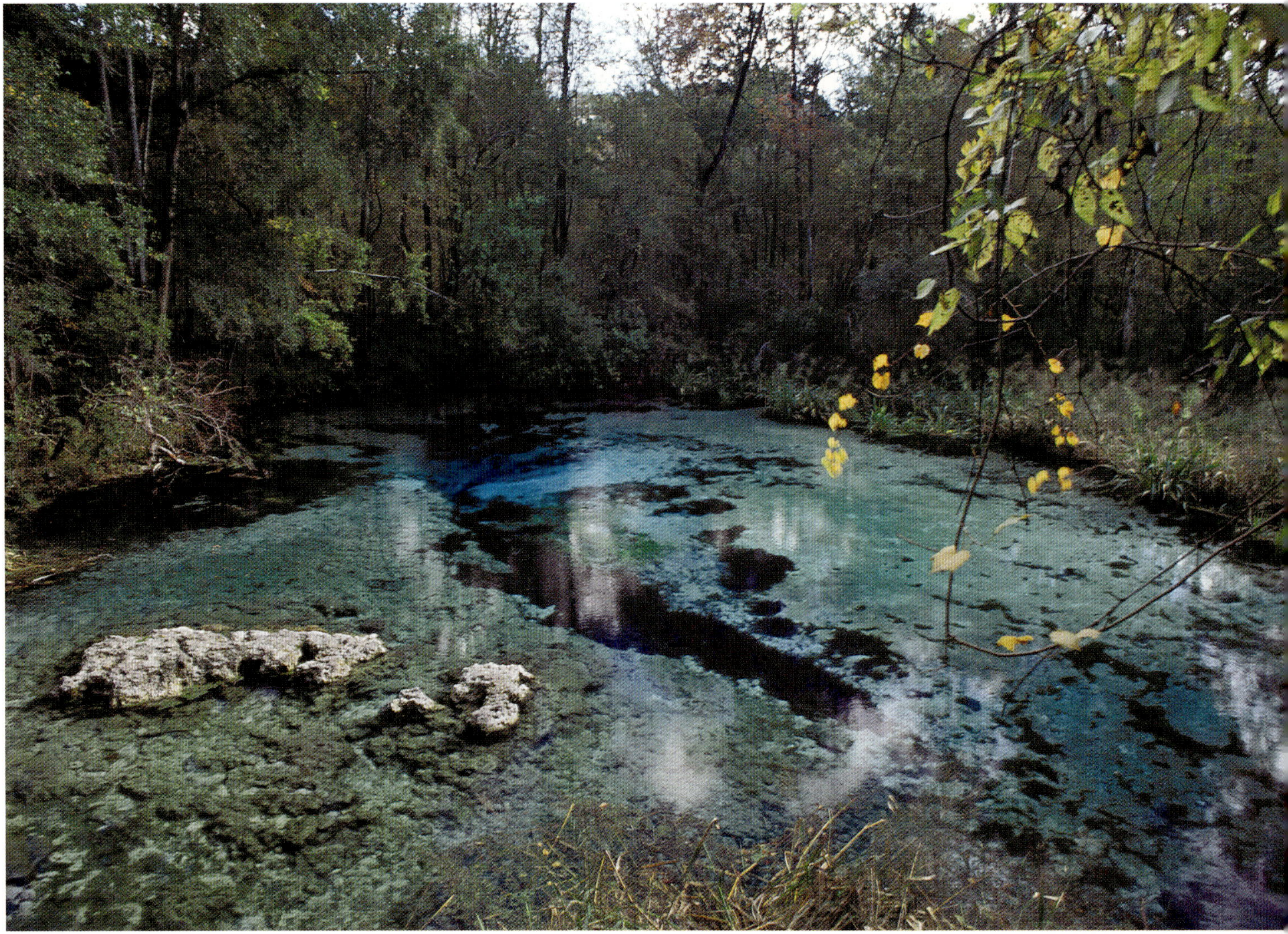

Ichetucknee

Sacred Waters

Steven Earl

Foreword by Susan Cerulean

14 13 12 11 10 09 6 5 4 3 2 1

Library of Congress Cataloging-in-Publication Data
Earl, Steven.
Ichetucknee: sacred waters/Steven Earl; foreword by Susan Cerulean.
p. cm.
ISBN 978-0-8130-3321-1 (alk. paper)
1. Ichetucknee River (Fla.)—Pictorial works. 2. Ichetucknee River (Fla.)—Description and travel. 3. Springs—Florida—Ichetucknee River Region. 4. Ichetucknee River (Fla.)—Environmental conditions. 5. Natural history—Florida—Ichetucknee River. 6. Ichetucknee River (Fla.)—Poetry. 7. Photography, Artistic. 8. Watercolor painting. I. Title.
F317.I24E23 2009
917.59'829–dc22 2008031702

The University Press of Florida is the scholarly publishing agency for the State University System of Florida, comprising Florida A&M University, Florida Atlantic University, Florida Gulf Coast University, Florida International University, Florida State University, New College of Florida, University of Central Florida, University of Florida, University of North Florida, University of South Florida, and University of West Florida.

University Press of Florida
15 Northwest 15th Street
Gainesville, FL 32611–2079
http://www.upf.com

In every walk with Nature one receives far more than he seeks.

John Muir

With deepest love, respect, and appreciation I dedicate this book to my parents, Leonora H. Earl and John F. Earl Sr., and to my dear friend, Crystal Earl, whose unique influences shape my being, making this unfurling of love possible.

To my beautiful children, Janaka, Noah, Enoch, Kamala, and Sam, to my family, and to my many sweet-souled and talented friends, I offer my heartfelt love and gratitude for your encouragement, support, and inspiration.

May this book honor and benefit the Ichetucknee River, while serving you, the reader, in connecting with nature and bringing joy to your heart.

Contents

Foreword

Susan Cerulean

An Ichetucknee float is one of the best ways I know to refresh my awareness of the vast palette of greens and blues offered on this earth. I crave slow time to contemplate stump-sunning turtles and meadows of white lilies, and the twirling of red maple seeds into the river's clear spring boil.

All of us deserve access to the wonders of the Florida landscape, most especially its unmatched rivers and springs. Long before computers and GPS units and spectrophotometers, our eyes, our ears, and our long-term memories kept us informed about the well-being of the places we lived. We are no longer the sharp-eyed hunter-gatherers who inhabited north Florida for twenty thousand or more years or the sweat-soaked naturalists who also approached the Ichetucknee River, reverently, on foot. Present-day pilgrimages usually involve air-conditioned automobile travel and small mountains of improbably knotted inner tubes on our car tops. Even so, our senses light up powerfully when we approach this tiny paradisiacal run.

Studies amassed by educator Joseph Chilton Pearce are showing that our indoor, electronically based lives are significantly dulling our brains' aesthetic capabilities. Fifteen years ago, people could distinguish 300,000 sounds; today, many children can't go beyond 100,000, and the average is 180,000. Twenty years ago, the average subject could detect 350 shades of a particular color; today the number is 130. Researcher Marcia Mikulak found that a child in a technologically advanced society has developed 20 to 25 percent less sensory awareness than the preliterate, or "primitive," child in the jungles of Brazil, Guatemala, and Africa.

So it's a happy paradox that the number of recreationists, spinning and plunging in their tubes from river rise to takeout, continues to increase. But they are taking their toll on the Ichetucknee River. The famed Florida naturalist Archie Carr once wrote: "[The tubers are] committing no violence on anything—just never ceasing to go by. . . . The magic of the loveliest waterway in the state is being eroded by the joyous, innocent, incessant passing of people on old inner tubes."

The river's native animals, emerged from egg or womb into this bright-lit heaven, must learn to contend with the unceasing pulses of human visitors or leave. The otter survives our incursions by slipping and breathing quietly along blown-down trees and at the edges of eelgrass meadows; the beavers that chew bark from the bases of trees must take their meals by night. Today, I float not five feet from two wood ducks tucked up into themselves, resting one-legged on a log. They watch me through a single open eye, chancing that they will go unseen, perhaps noticing how often humans are fixed on what is to come around the bend, rather than what the present moment offers here and now. Other creatures haven't survived the human fascination with the river and have been largely plucked from its bed or driven out by declining water quality. These creatures include alligator snapping turtles, rare snails, and translucent spring crayfish.

The Florida Park Service is working to protect the river by fencing off the most sensitive spring tributaries, setting caps on the numbers of tubers allowed on a given day, handpicking exotic plants from the river's edge, and generally managing the clamor for access. But threats to the river are intensifying, still human-derived but much less visible and not so easily measured by eye.

I study the gentle floodplain sliding past my kayak and look deeply into the forested terraces that hold this river like a lover. These uplands remind me of the river's trace, a dry valley that overlies a major underground conduit extending south and west all the way from Lake City and that supplies most of the river's clear water. Whatever is put on the trace's surface will eventually make its way to these waters; at present, limestone mines, septic tanks, and fertilizer runoff are some of the human activities that threaten the integrity of the river. The Ichetucknee needs our urgent advocacy.

To that end, Steven Earl has artfully constructed this collection of photographic and watercolor images and poetry, with contributions from Cathy Nagler and Crystal Earl. Here, he celebrates the Ichetucknee's singular beauty, hoping to open wider the eyes of all who love her. Earl, an artist and park ranger who once lived with his family within walking distance of the river's headspring, worries that the pace of our lives is desensitizing us to the natural world and with this work reminds us of the loveliness and the essential character of this beloved river. A slow perusal of this book, like a long float in the Ichetucknee itself, may allow us to emerge refreshed by the physical precision of crystal water and the emerald of eelgrass, ready to work on its behalf.

Susan Cerulean is a writer and editor living in Tallahassee. Her latest book is Tracking Desire: A Journey after Swallow-tailed Kites.

Introduction

Bill Hanrahan

One day's paddle on the Ichetucknee River changed my view of Florida; that the river still influences me on matters spiritual, philosophic, and poetic is tribute not only to a magical watercourse but also to the man who took me there.

Steve Earl listens to Coltrane and Santana. He cheers for the Gators. He drinks green tea and eats vegetarian. He loves percussion. He makes instruments out of bamboo and can turn a box of toothpicks into a maraca. He's a proud father, an inspired painter, and a stunning photographer. But more than anything else, the thing that makes Steve whole is that he breathes the natural world. It is part of the peace he exudes when you are with him. It's a spiritual thing, a cosmic connection with trees, plants, birds, and water. Most amazingly, it is something he shares.

My wife, Ellie, and I met Steve about a year before that sunny December day when we slipped two canoes into the calm riffles near the headwaters of the Ichetucknee. We were doing volunteer work for the Florida Park Service, and Steve was the ranger assigned to oversee our stay at Gold Head Branch State Park. Steve exuded the best qualities of a ranger. He cared for the right reasons. He knew every inch of his park, and his powers of observation stretched well beyond my own. Once, when I brought his attention to the cacophony of frogs coming from a nearby lake at dusk, Steve listened in silence for a moment, then asked Ellie and me how many different kinds of croaks we could decipher from the chaos. We spent the next twenty minutes "seeing" with our ears.

Steve shared with us his belief in the wisdom of Native American Indians and their respect for the earth. He would marvel at the spiral on the top of a seashell and proclaim it the "Holy Organic Pattern," as dubbed by his friend Johnny Dame. He could share a laugh by the campfire or just as easily sit in silence, transfixed by dancing flames and the music of crackling wood. He loaned me a great book on John Muir and told me to pass it on when I finished. One evening, after we had shared a veggie stir-fry dinner, this modest and unassuming man revealed to us one of the most dynamic aspects of his personality—his paintings.

Flipping through copies of some of Steve's watercolors, I realized just how little we sometimes know of a person, how far beyond our initial perceptions the veins of talent and inspiration can run: violet depths and turquoise swirls, sparkles and splashes bursting off the canvas, the pink and yellow hues of sky over the greens of the land, rich blues spilling out of the earth, a swallow-tailed kite soaring overhead. Each painting was unique, each an impressionistic burst of originality, at times dreamlike, at times mystical, and each one a variation on the same theme—Steve's river, the Ichetucknee.

Another night, at the state park, Steve asked if we'd like to see some of the slides he'd taken, photographs of a most special place. We gathered with other campers in a community room. Steve stood before us, and as he focused the slide projector and prepared accompanying music, he asked the group to please refrain from talk or questions during the show. He wanted the slides and the music to set a mood. He urged us to relax, enjoy the show, and let the river take us.

The heart of Steve's slide show—and the essence of his creativity on canvas—is what you now hold. This beautiful book is the work of a man who cares deeply, a man whose passions for photography and art point toward a spiritual unity with nature.

When Steve took Ellie and me to the Ichetucknee on that sparkling winter day, we entered the world you see in this book. For three hours we encountered no other people but ourselves reflected with the natural surroundings on the water. Humans, we're told, are made of some 80 percent water, and on the Ichetucknee you believe it. The river fosters not a "connection" with nature but inclusion. We are the earth, and it is beautiful.

The love and talent Steve poured into this book make it pure hydration for the thirsty soul. Open it. Turn the pages slowly. Drift, dream. Let it flow . . .

Bill Hanrahan is a freelance writer and former veteran Connecticut newspaper reporter. He and his wife, Ellie, have been traveling the United States for the past eight years, in search of America's wild places.

Preface and Notes

Nothing had prepared me for the day in 1983 when I first met the Ichetucknee River. Not the influences of a mother with an open, tender heart, a wise soul who was a disciplined meditator and lover of God. Not the influence of a father who was a nature photographer, jazz musician, artist, naturalist, and environmental champion.

Not the experiences of long-ago vacations spent in spectacular places such as Georgia's wild Guale Isles, the mountains of Appalachia, or the unspoiled coastline of Florida's panhandle.

Not even the stories told by my father of having canoed down the Ichetucknee some ten years prior, while working on his photographic book, *John Muir's Longest Walk*.

No, nothing had prepared me.

Here I sit on the bench at Blue Hole, shivering, stunned and in awe, after swimming out over the boil and peering down into the eye of the spring. I have no previous knowledge that the phenomena of springs and spring-fed riverine systems even exist. This is all totally miraculous.

Shimmering shafts of sunbeams streak down the turquoise depths, the play of light and shadow dancing on the rippled white sands below. Unseen fossils of mastodon, American lion, and giant land sloth lie asleep in her watery womb. Timucuan-knapped spear points lie buried deep in her banks of rich floodplain earth, unseen by human eyes for thousands of seasons.

Jeweled iridescent dragonflies of every hue cruise in and out of stately cypress trees, some alighting momentarily atop brilliant red cardinal flowers and ivory white spider lilies.

Above the cypress tree sisters I sit admiring, sail several of the most magnificent bird in all of God's creation: the swallow-tailed kite, with its stunning black and white plumage and deeply forked tail. In pursuit of dragonfly, they put on an aerial show like no other, executing maneuvers that boggle my mind and make my heart soar. I am spellbound by this never-before-seen creature of overwhelming beauty and grace.

A mixture of subtle scents wafts in the air, including cypress oils, spring water, floodplain humus, and at water's edge, the delicate fragrance from eelgrass blossoms as they periodically bob to the surface on swaying strands.

Sitting quietly and counting my new-found blessings, I become aware of the mind-centering drone of summer cicadas, fresh from their years-long sojourn beneath the earth. As does ancient music, this droning has the effect of freeing my mind of thought and thus opens my consciousness to receive the gifts of my surroundings. Relaxing deeper, I begin to feel at one with the ebb and flow of my realm: the swirling water, the thrum of insects, and the life force within me. Unbeknownst to me, on a much deeper and subtler level of my being, a life-altering transformation is taking root; I am bonding with what will become my home and beginning to recognize what is known as "sense of place." I have found the center of the world.

For the next twelve years I lived close to the land in full-time relationship with the river. A primitive log cabin over a hundred years old, atop the underground headspring source in the river trace, provided shelter and hearth.

Living in such close proximity to the river, I was able to spend a vast amount of time interacting with the environment. At first this included furthering my sense of place by practicing the art of open-eyed meditation in remote locations, through which I became keenly aware of my surroundings and the interconnectedness of all things, absorbing these experiences into the deepest level of my being. Out of this connectedness arose many other things: peace, knowledge, gratitude, artistic inspiration, and stewardship of place.

The Ichetucknee became my muse, the focus of my photography, filmmaking, painting, and musical composing. I observed, too, that the very act of composing through the camera lens all day further focused my meditation and my insight into this most beautiful place. It is my prayer and hope that my life and work will inspire people to devote space in their busy lives to slow down, be still, be aware, and commune with nature regularly. I especially hope that parents and educators will reunite children with the wonders of nature. Positive environmental and personal change occurs each time a person bonds with the natural world, thus fostering peace within and a desire for stewardship of the planet we call home.

Drifting to sleep in the cabin, I often thought of the underground river beneath me and the untold gallons of crystal-clear water forever rushing to the headsprings and the light above. The Ichetucknee system (like the entire planet) is compromised and threatened by human activity. Jim Stevenson, who has dedicated himself to monitoring and protecting the river and its basin, has written an essay for this book that explains these challenges and describes projects being undertaken to meet them. Please take the time to read his afterword. I encourage you to play a role in the stewardship of the Ichetucknee River by becoming actively involved with others who are standing strong in service.

Here's hoping we all enjoy the river and that we find delighted children splashing in pristine blue waters for millennia to come.

Notes on Illustrations/Paintings/Photographs/Poems

Illustrations

The Ichetucknee Springs Basin Map/Legend was lovingly rendered by friend and artist Johnny Dame. We collaborated on the concept and design. The components of the piece contain great significance, symbolism, and honoring.

The border's shape and incorporation of four directional circles were lovingly borrowed from a design that my friends Jim Valentine and Skip Lake conceived and rendered in the early 1970s for an exhibit map titled *The Land of Guale*. That map and accompanying exhibit were used to celebrate and support the release of Jim's and my father's photographic book, *Guale, the Golden Coast of Georgia*. That book continues to this day to help protect many of Georgia's wild coastal islands. My father, Jim, and Skip were early photographic and artistic mentors, respectively, and this inclusion was among various attempts to pay tribute to those influences.

The use of the circles in the map's design is a means of showing respect for the Native American belief in honoring the planet as a whole, by acknowledging the four directions. Each of the circles supports one of four kingdoms: human, animal, plant, and mineral. Each representative tells a story of the river and my family's life there. The season of the map is spring, chosen to perpetuate regeneration and hope for the future. Our children depicted likewise represent future generations receiving the blessings of the river. The geometric pattern within the border was respectfully reproduced from a Timucuan water vessel recovered from the river.

The area of the map within the heart is the upland basin that is drained to Ichetucknee Springs/River. Obviously, this is the area of the most critical concern for contamination, and therefore it is imperative that it be managed wisely and lovingly!

The colors used within the map area represent elevations above sea level. Blue = 10–25', Green = 25–50', Tan = 50–75', Brown = 75–100', and Purple = 100' and above.

Paintings

The paintings of the river and springs in this book are a sample of a body of work compiled over a twenty-three-year period. All paintings are watercolors, many of which are mixed with the waters from the scene. Very little on-site painting is done. Rather, I will spend hours at the scene absorbing the surroundings into my being, collect water, and return to the studio to allow the day to flow through me onto paper. In addition to applying paint with standard brushes, quite often I use plant matter from the scene such as Spanish moss, reindeer moss, and old man's beard lichen.

Photographs

The majority of the photographs in this book were taken with Nikon 35mm cameras and Nikkor lenses. The film used was Fujifilm Velvia 50, almost exclusively. A very small number of photographs herein were produced from digital cameras. Many of the photographs in this book would not have taken place had it not been for the third major photography mentor in my life: renowned Florida nature photographer John Moran. Most of the years my family and I lived at the river were very lean. Knowing this and fostering my potential as a photographer, John would drop by the cabin once or twice a month and gift me with rolls of the finest slide film available and share important techniques, often taking me with him on outings to the river for hands-on lessons and experience. Thank you, John.

The photos and paintings in this book are arranged in descending order from north to south, starting with looks into karst openings in the Ichetucknee trace and proceeding down the river to its confluence with the Santa Fe River. Along the way, side trips on foot and by water reveal the wonders of springs, the recesses of the floodplains, and even the upland sandhill ecosystem, bordering the river on both sides.

Poems

Four of the six poems in this book are from a larger collection of Ichetucknee poetry by poet Cathy Nagler. The other two poems were written by Crystal Earl and myself.

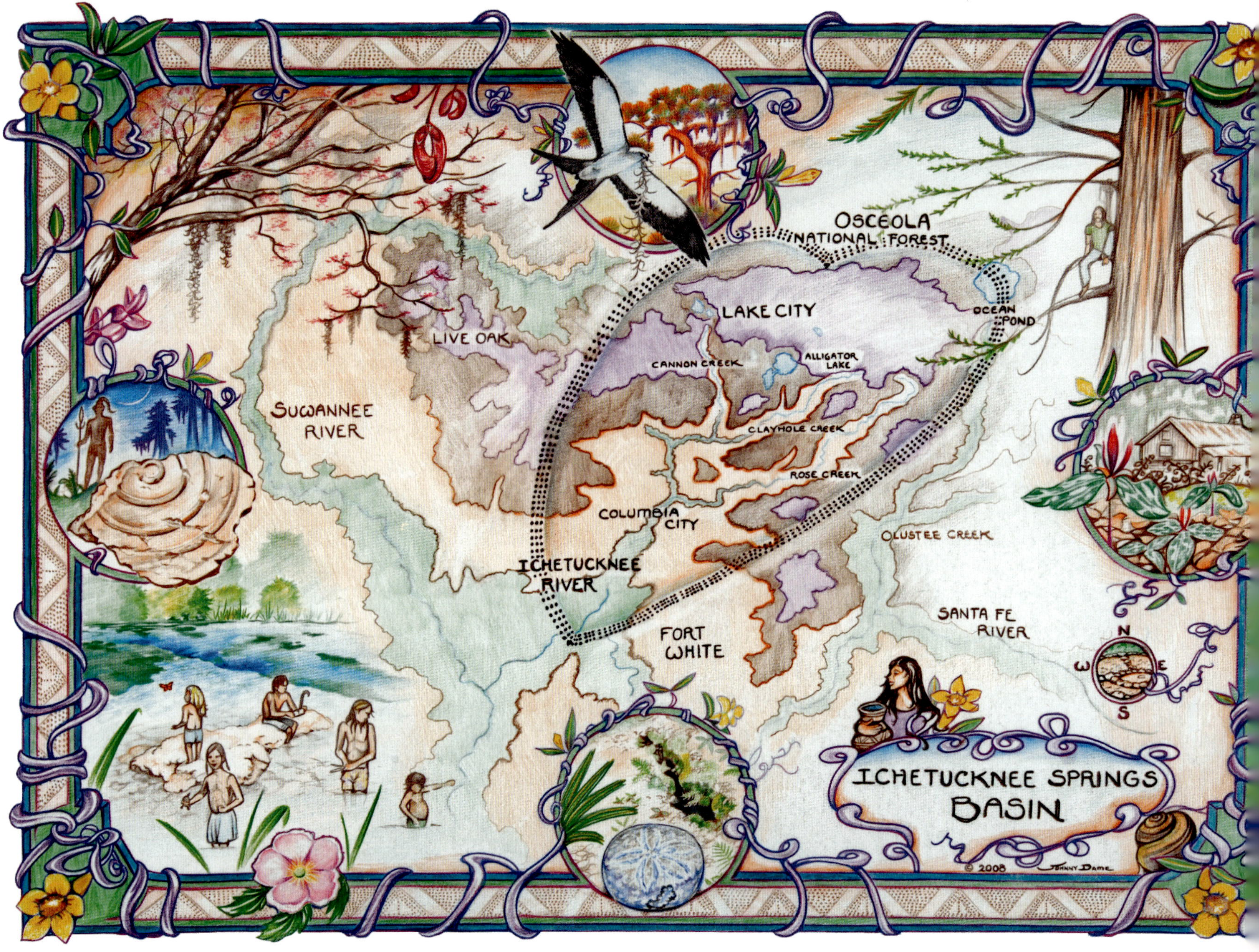
OSCEOLA
NATIONAL FOREST
LAKE CITY
OCEAN
POND
LIVE OAK
CANNON CREEK
ALLIGATOR
LAKE
SUWANNEE
RIVER
CLAYHOLE CREEK
ROSE CREEK
COLUMBIA
CITY
OLUSTEE CREEK
ICHETUCKNEE
RIVER
SANTA FE
RIVER
FORT
WHITE
N
W
E
S
ICHETUCKNEE SPRINGS
BASIN
© 2008
JOHNNY DAME

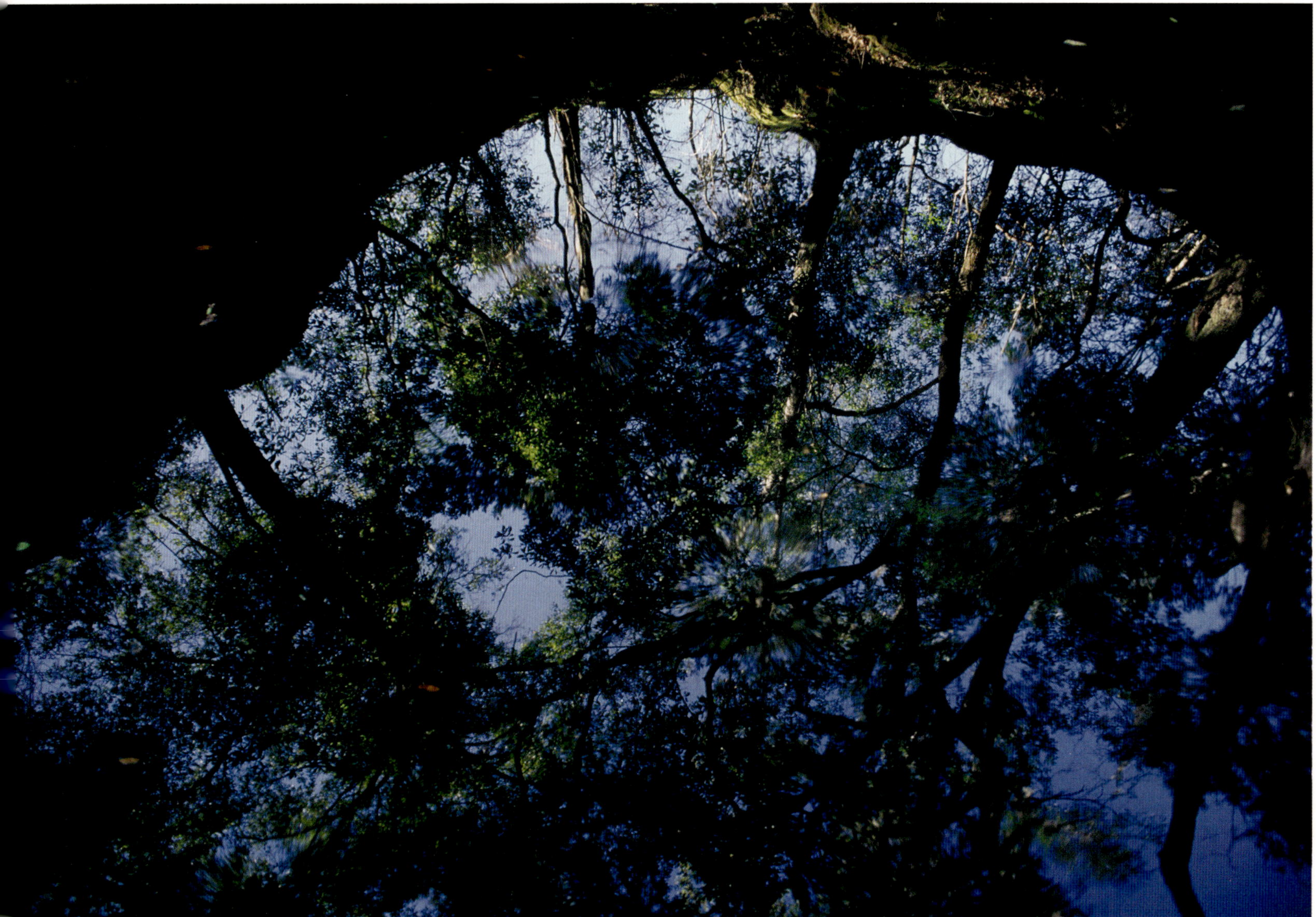

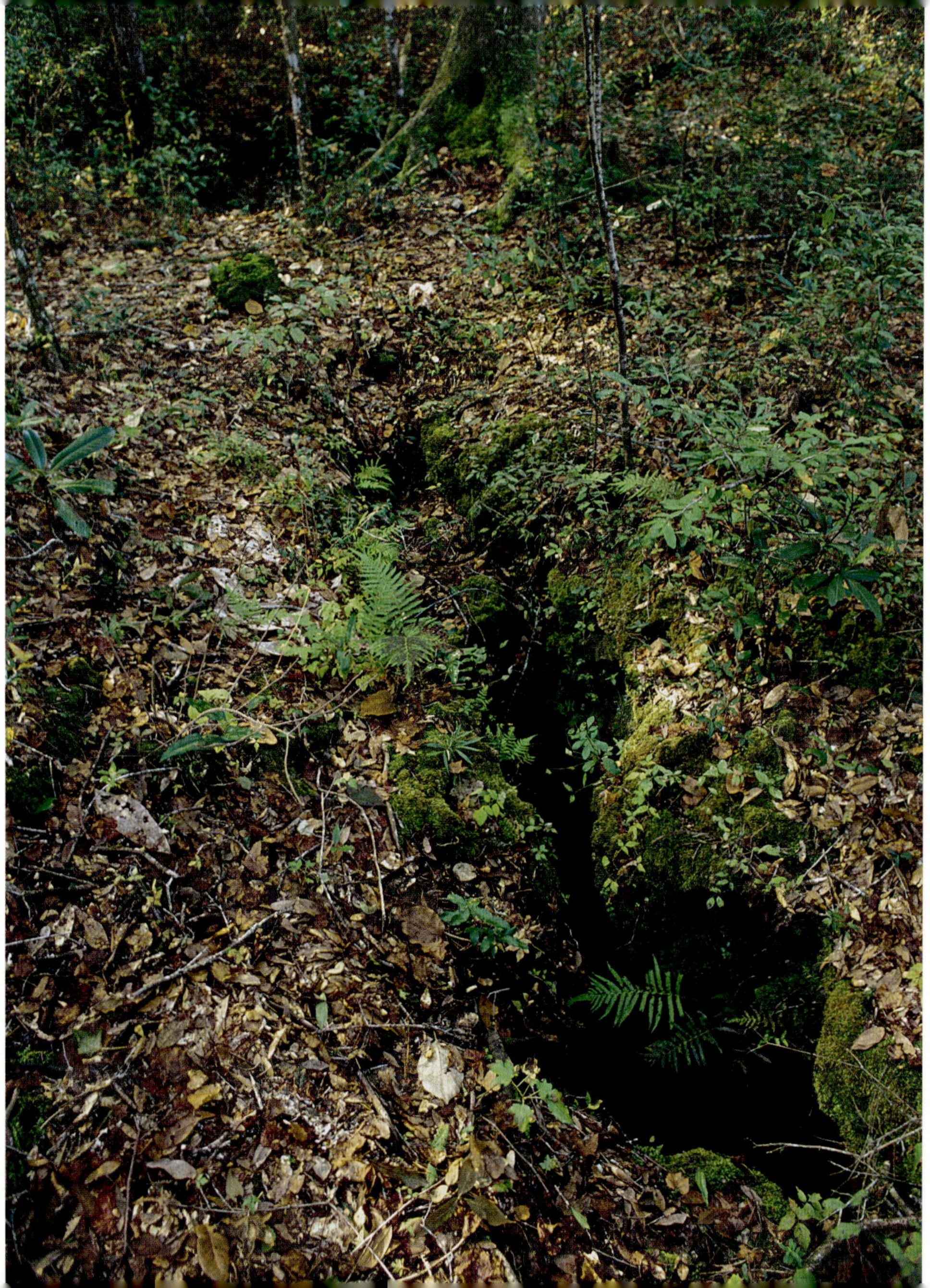

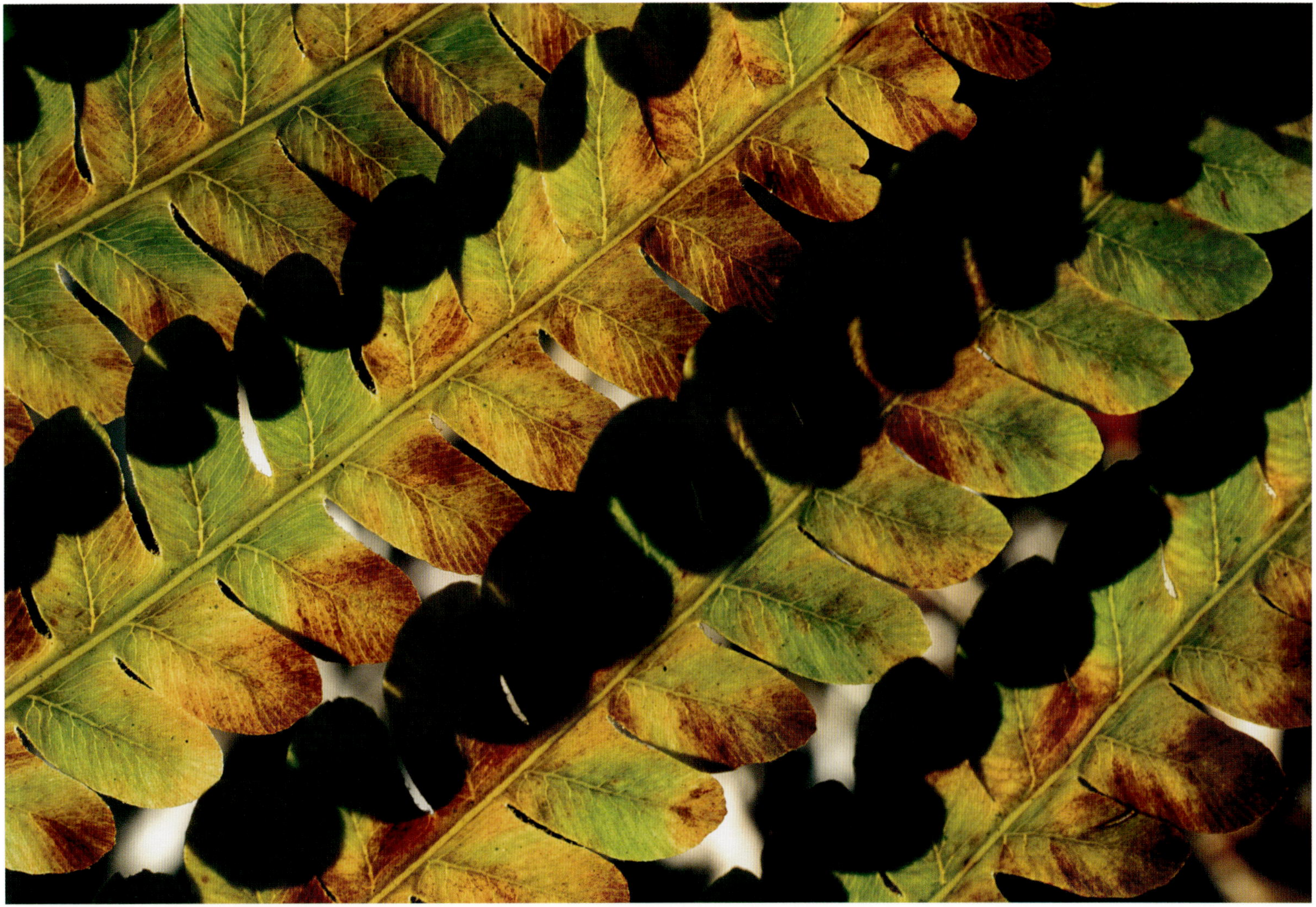

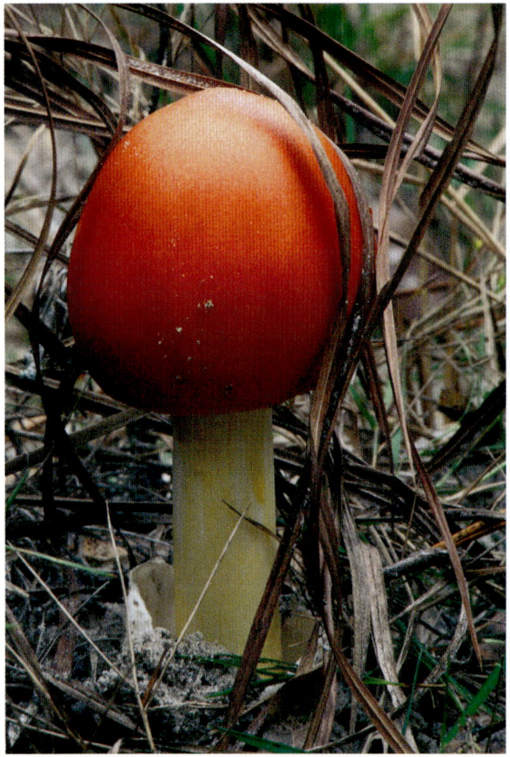

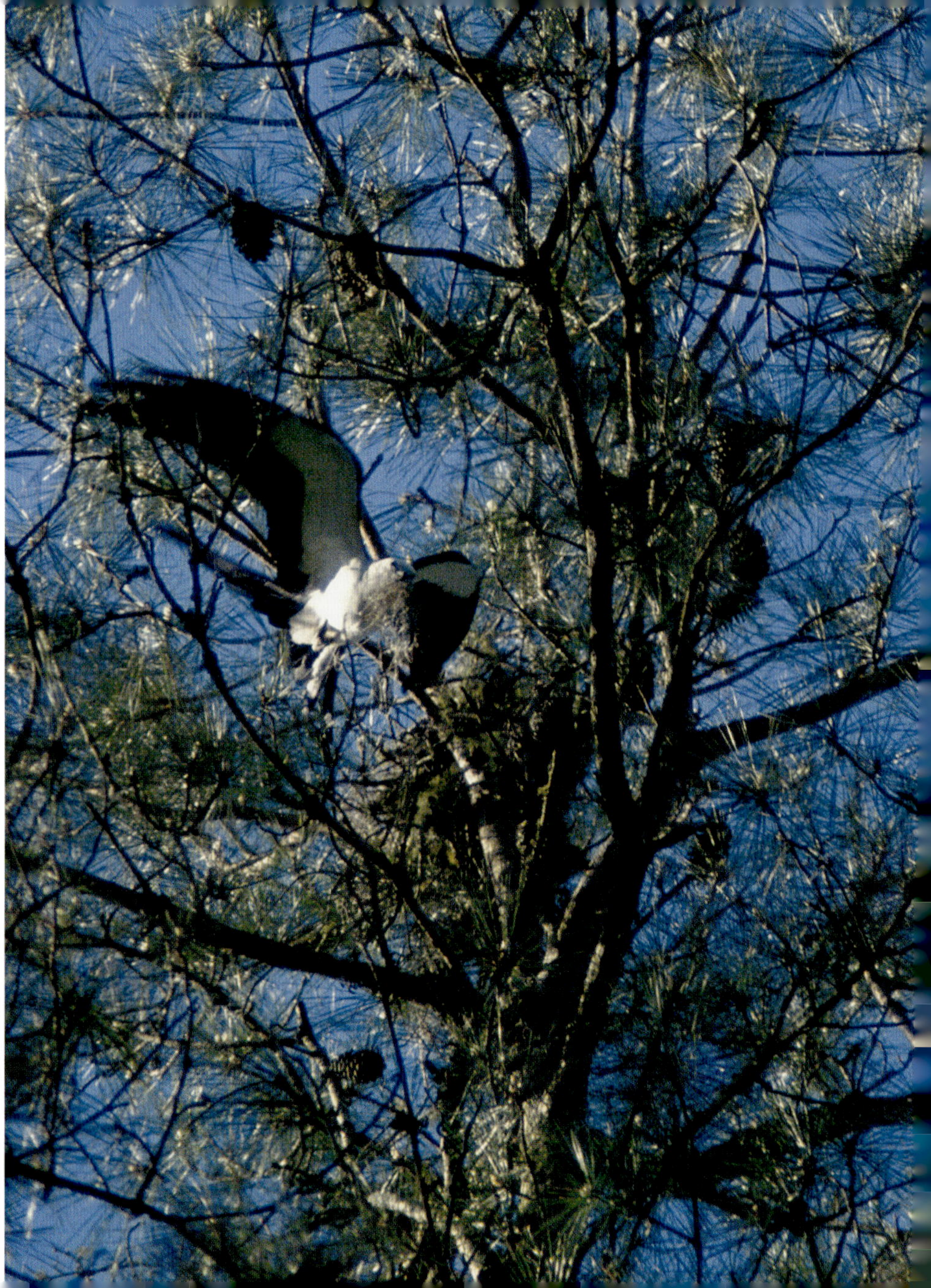

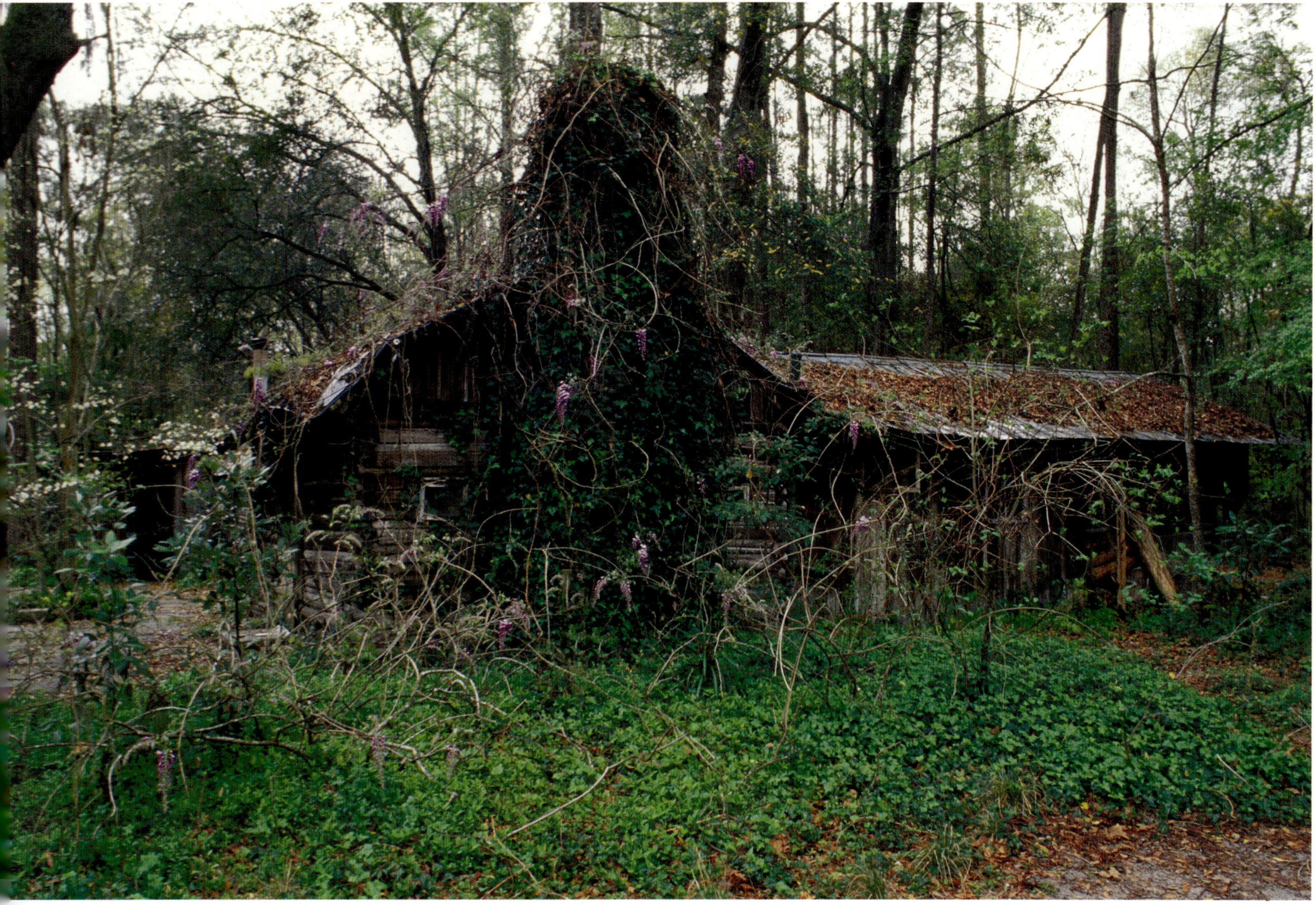

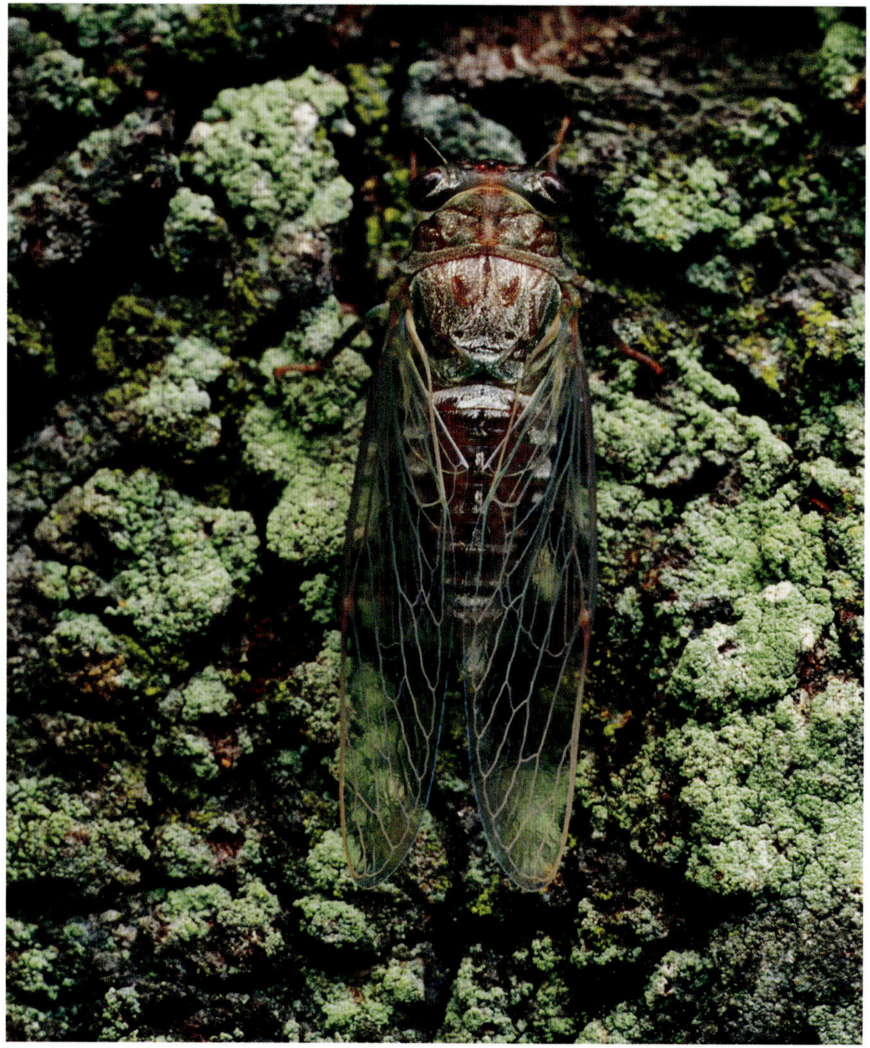

winged beauty at rest
motionless
but for intermittent
rise and fall of male droning

a cued chorus
building in sound and duration
with the increasing heat and humidity
of the summer day

a crescendo
climaxing into a steady hum
working its spell
from background into foreground

inducing a trancelike state
thoughts becoming wordless
with the constant whirring thrum

nature's meditative powers
transport to the center of being
succumbed to a state of silence
stillness in the sound

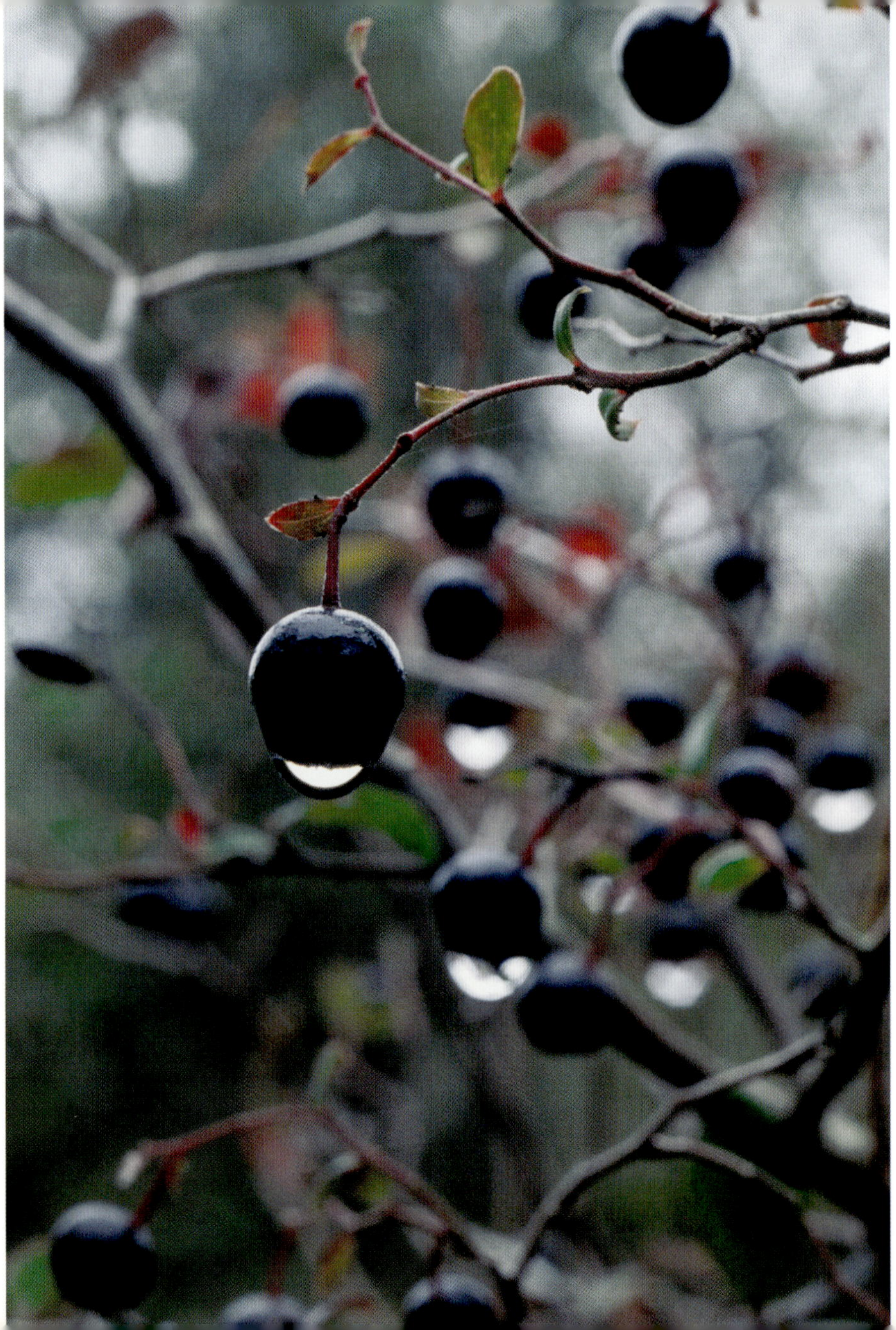

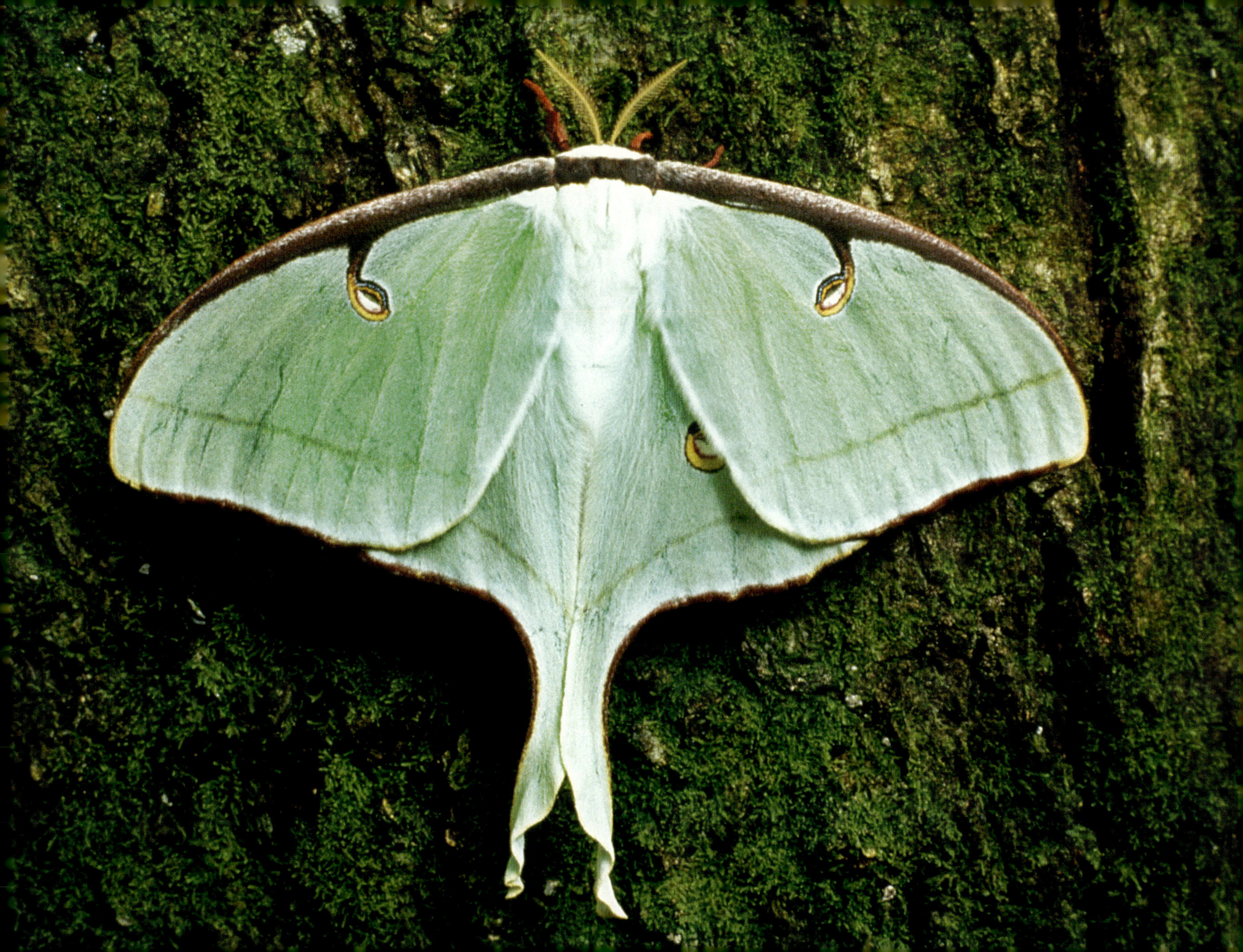

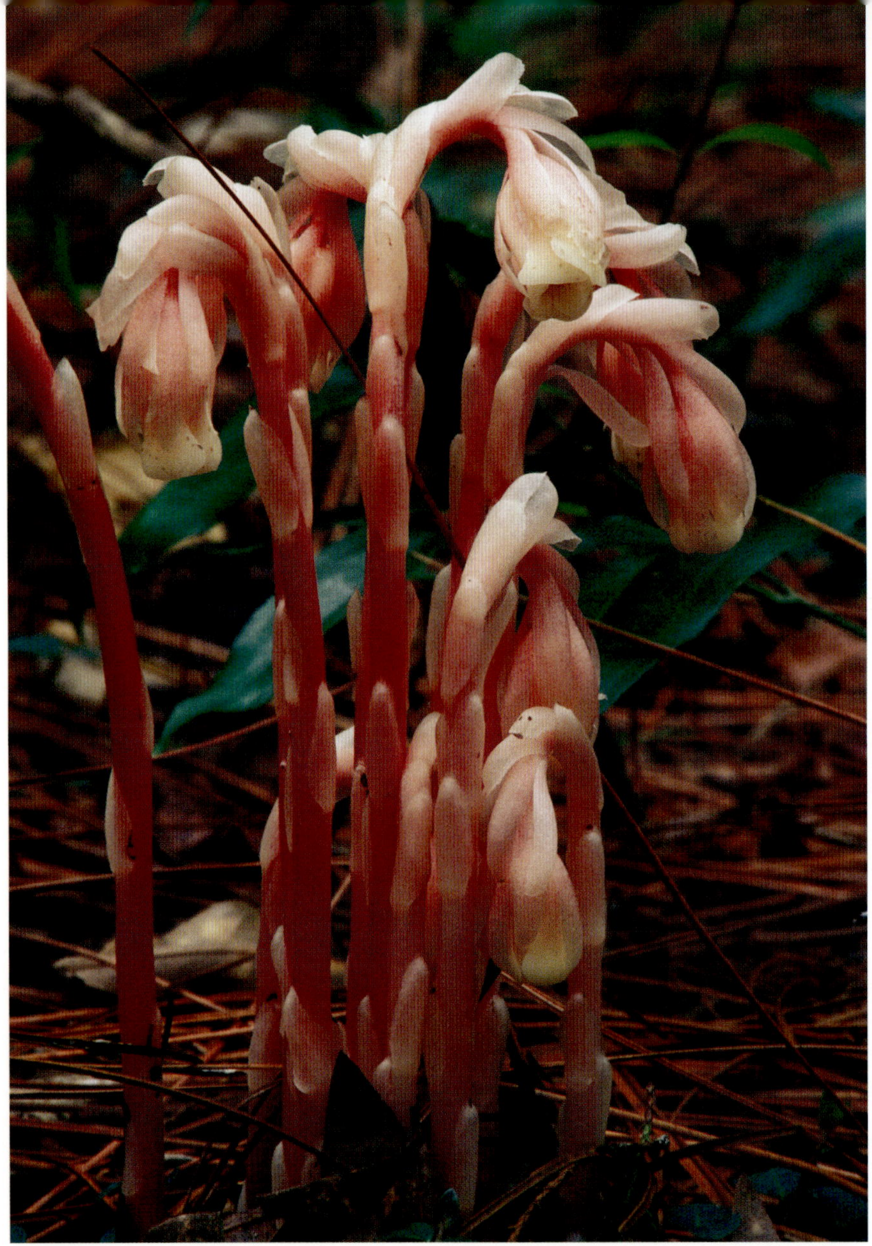

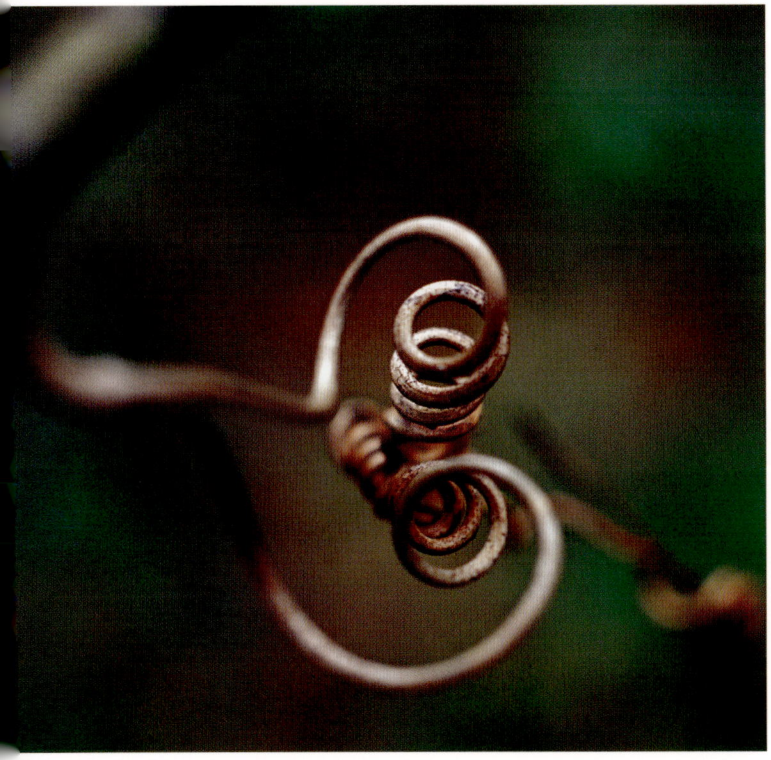

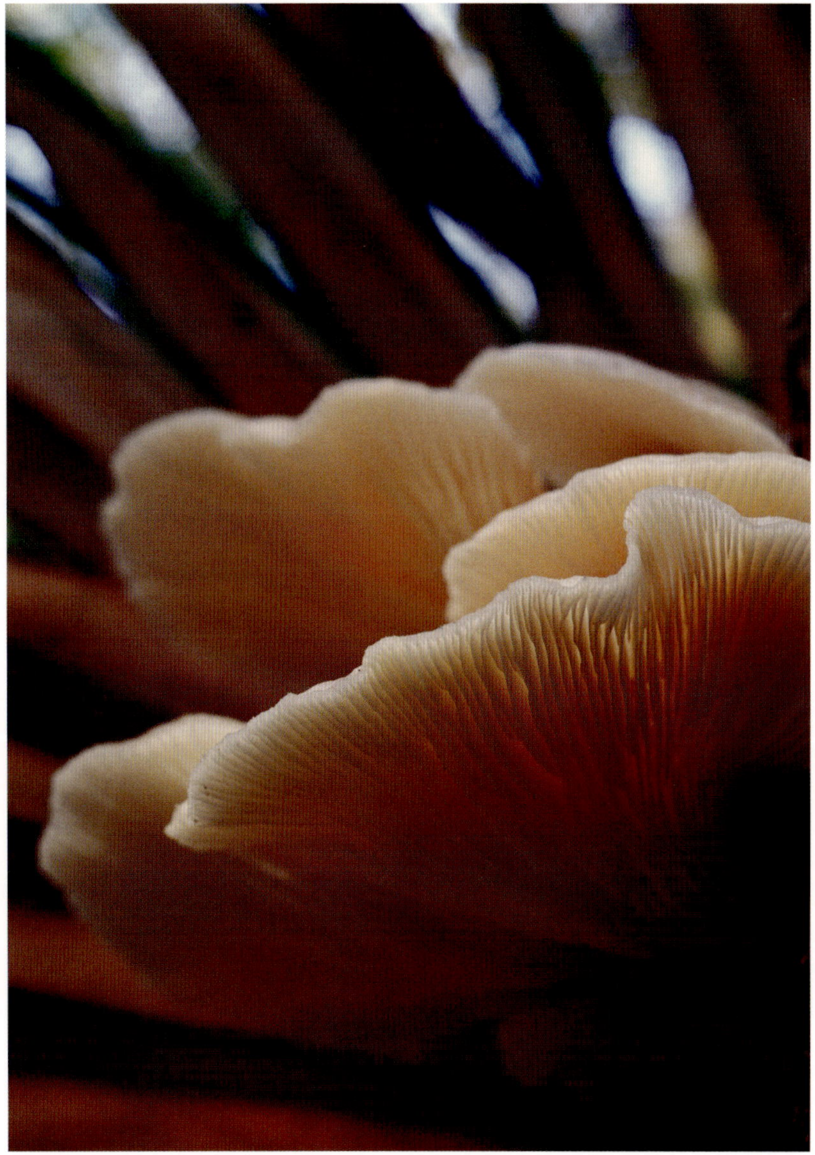

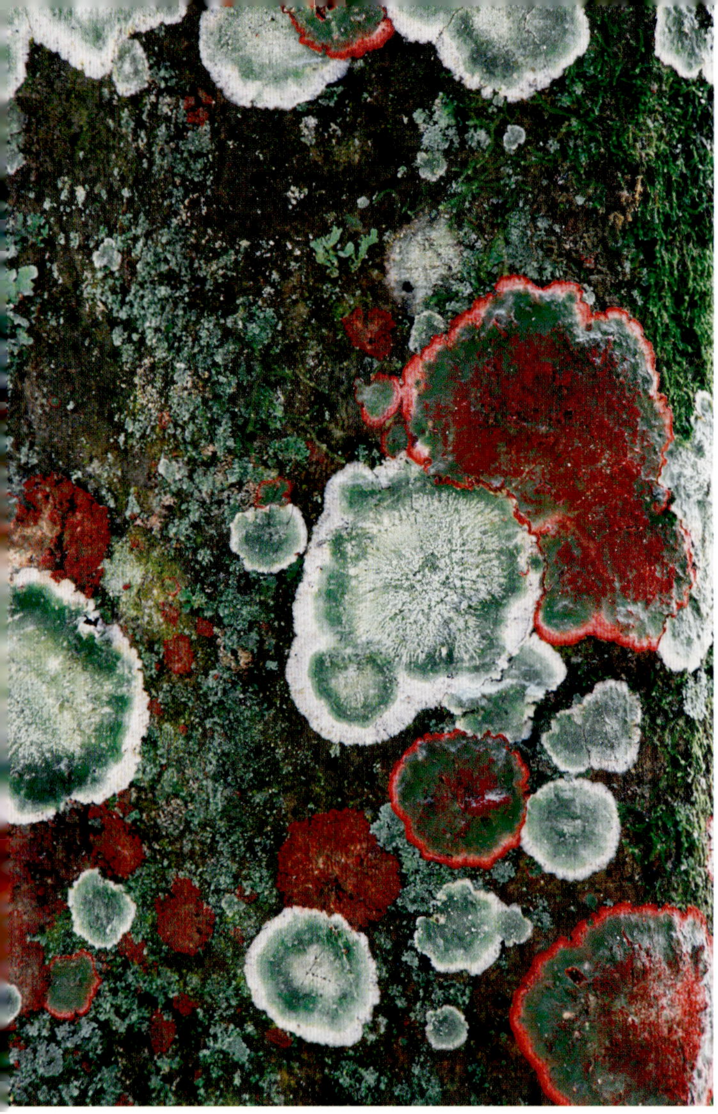

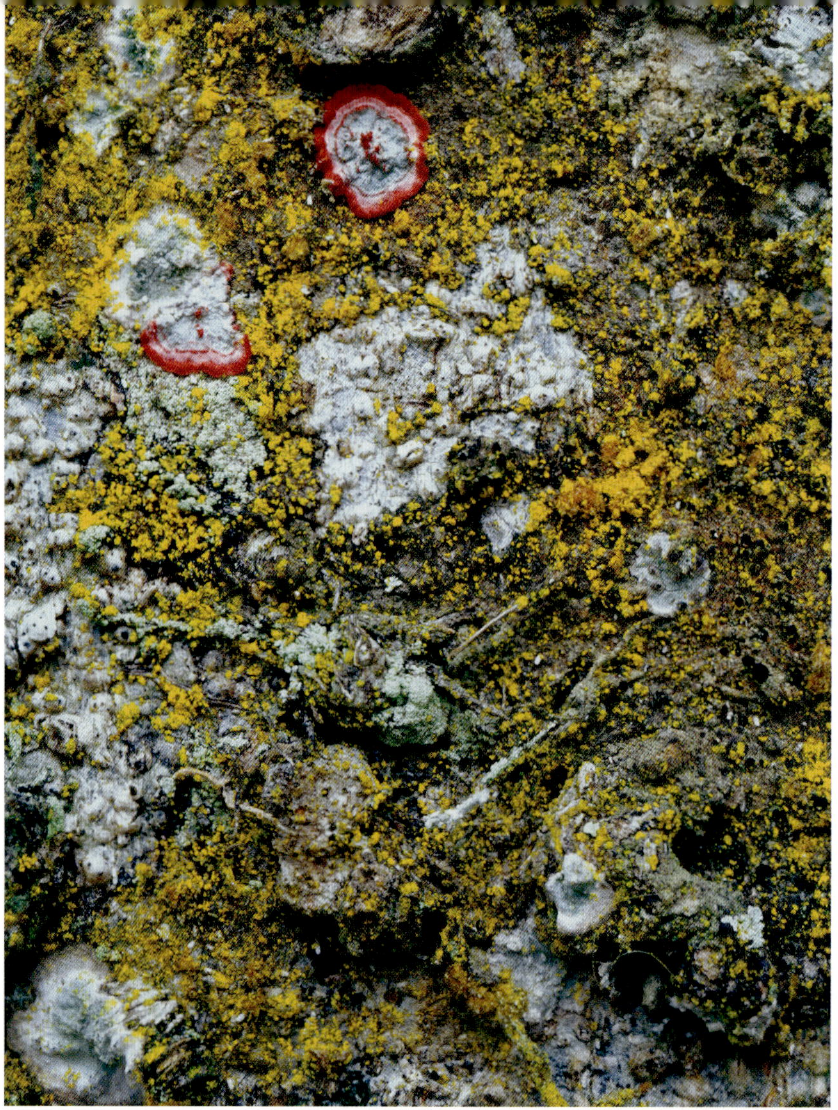

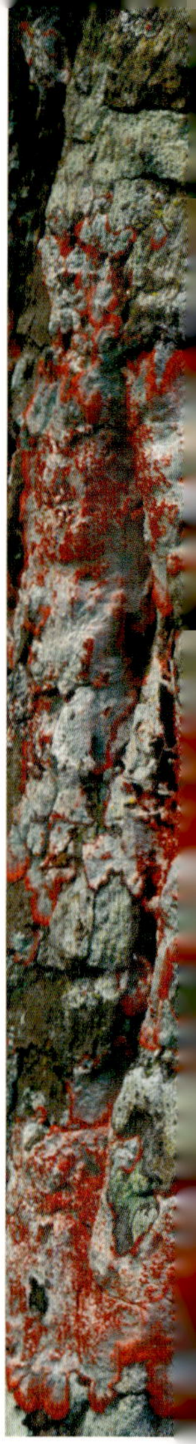

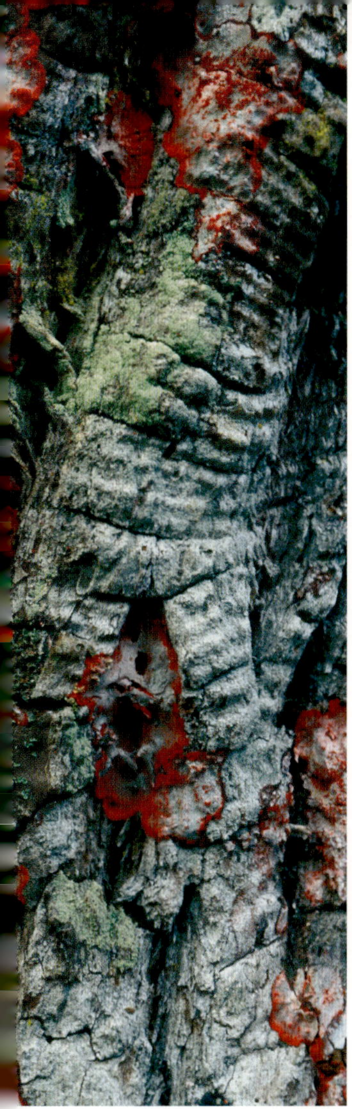

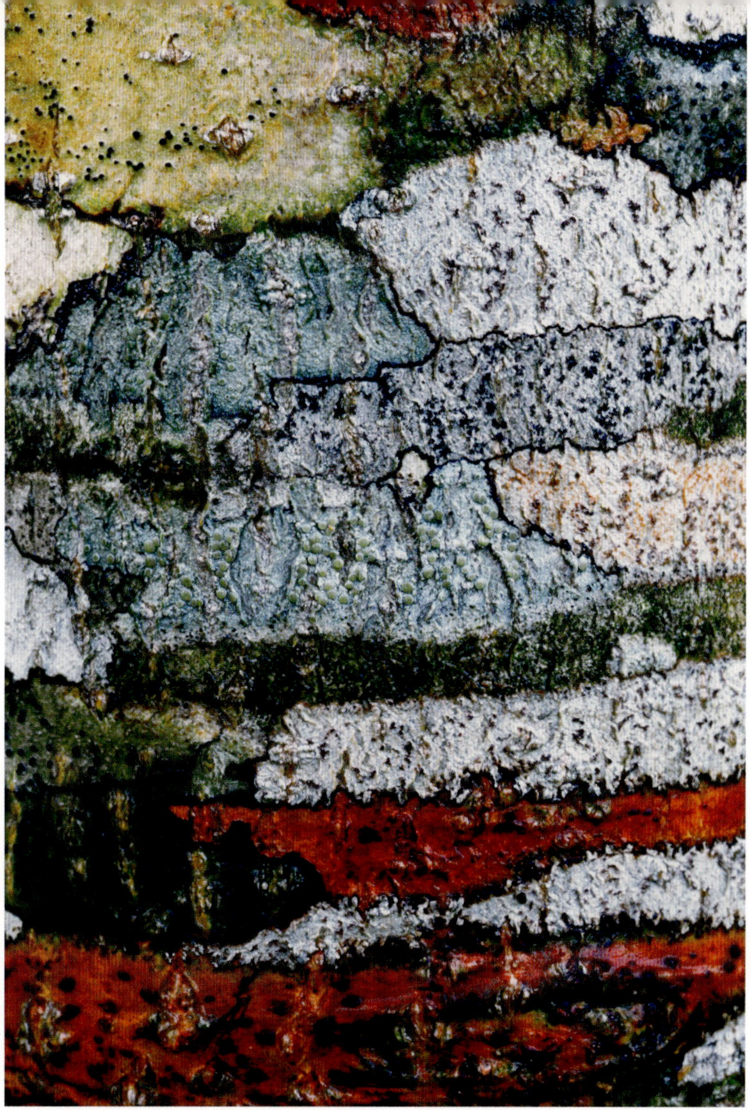

a mesmerizing patchwork of
lichens
an amazing array of color and pattern

a dazzling vision of and from a union
algae and fungi joined in life-giving
synthesis
uniquely revealed in divinely inspired
art
aware eyes fill with wonder

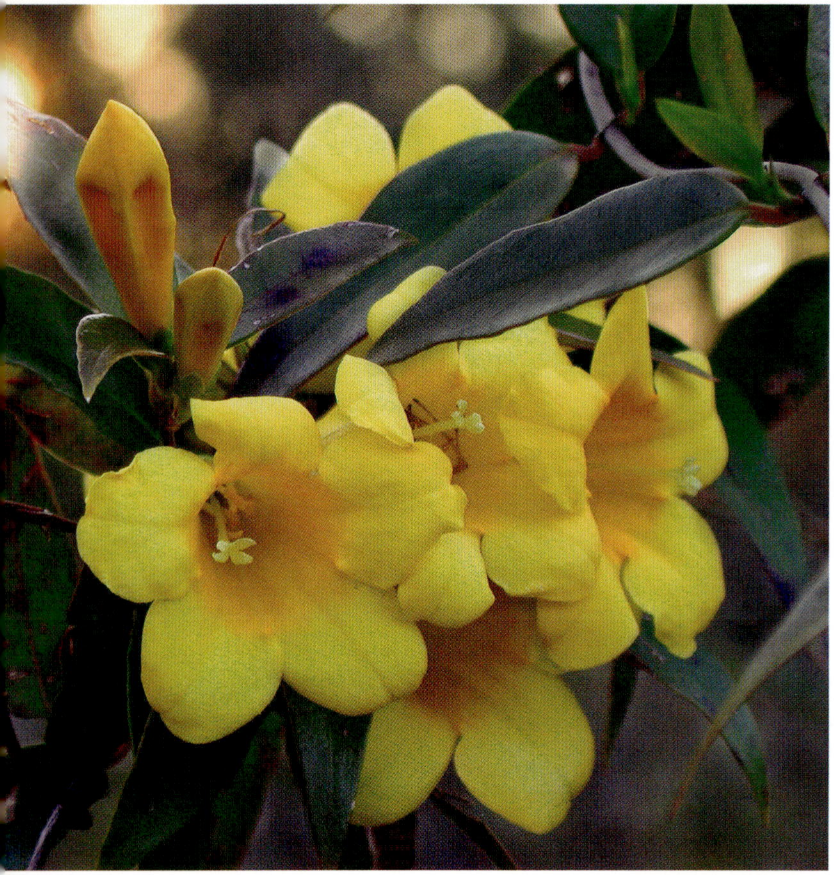

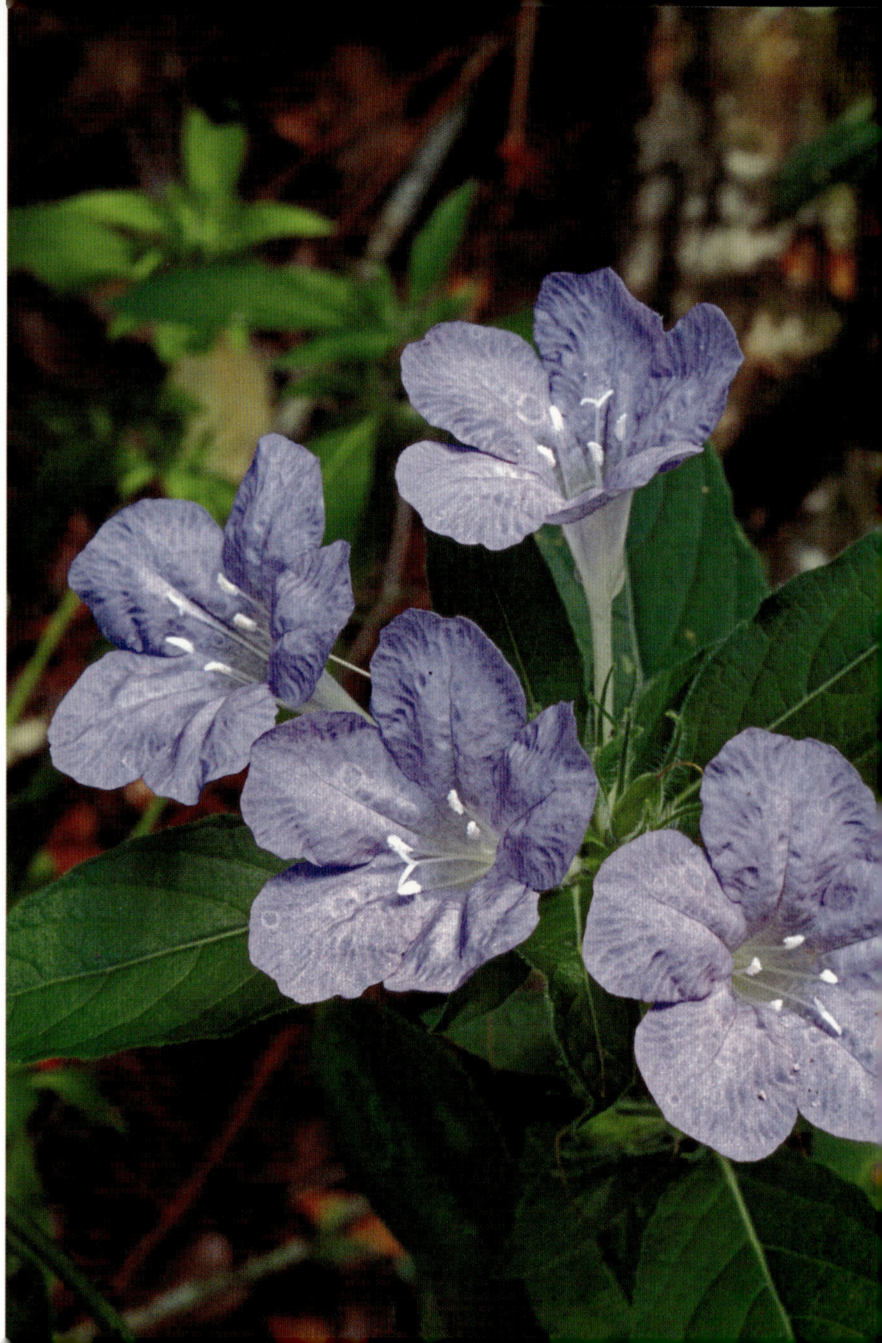

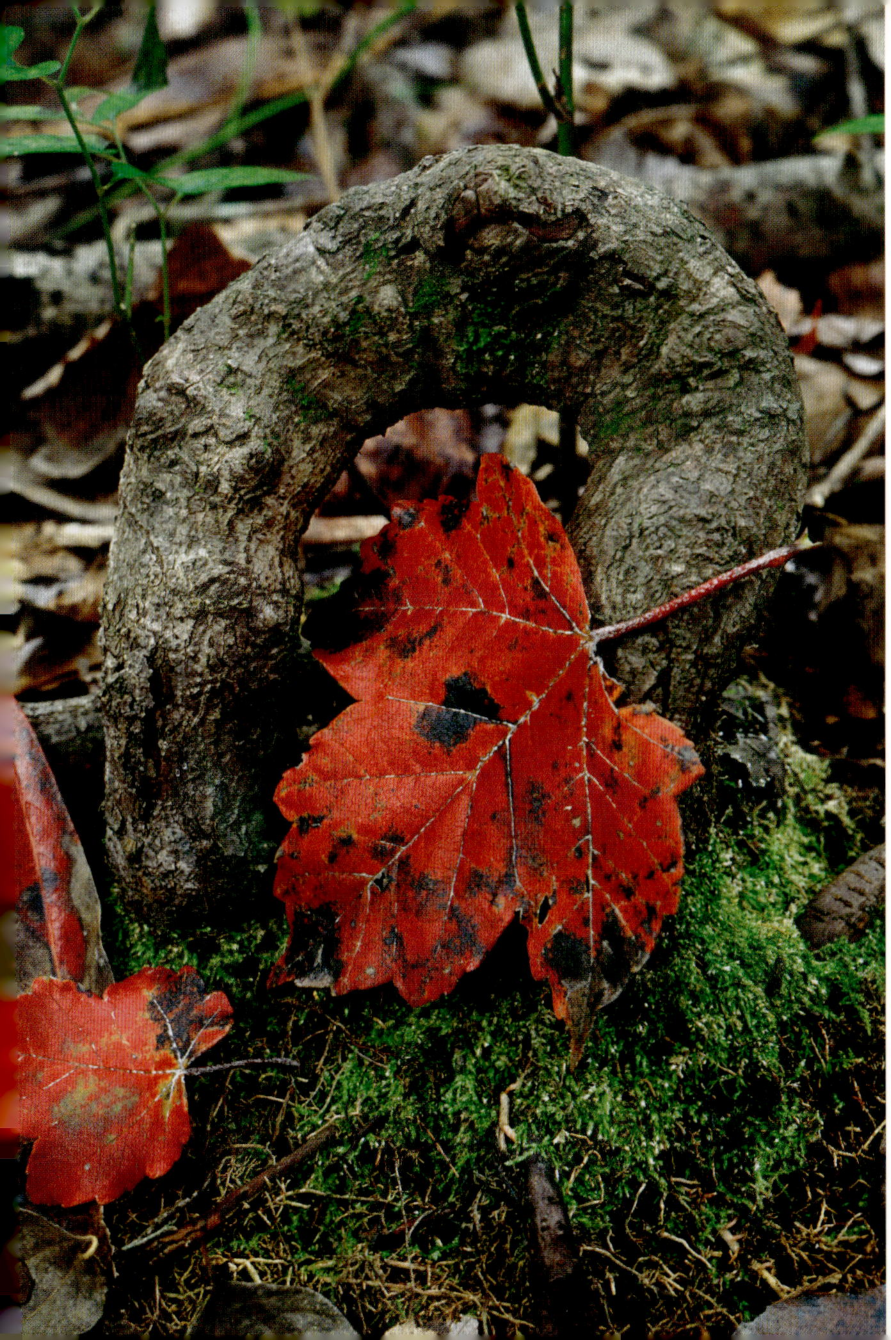

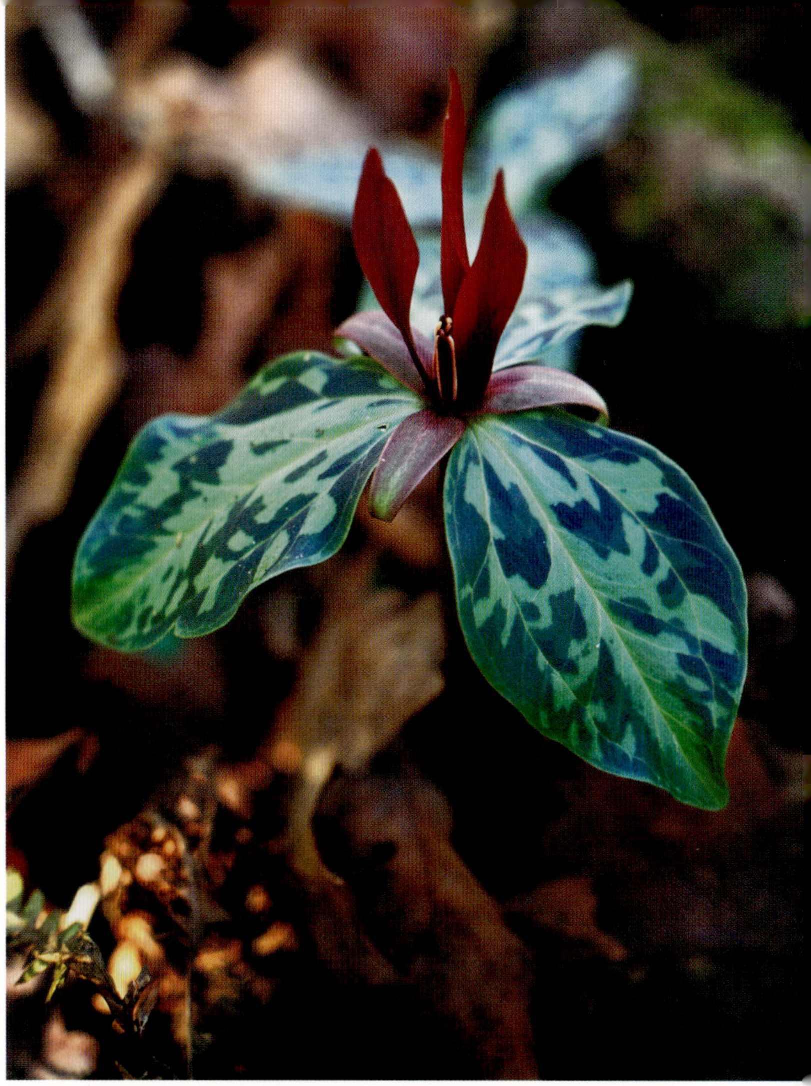

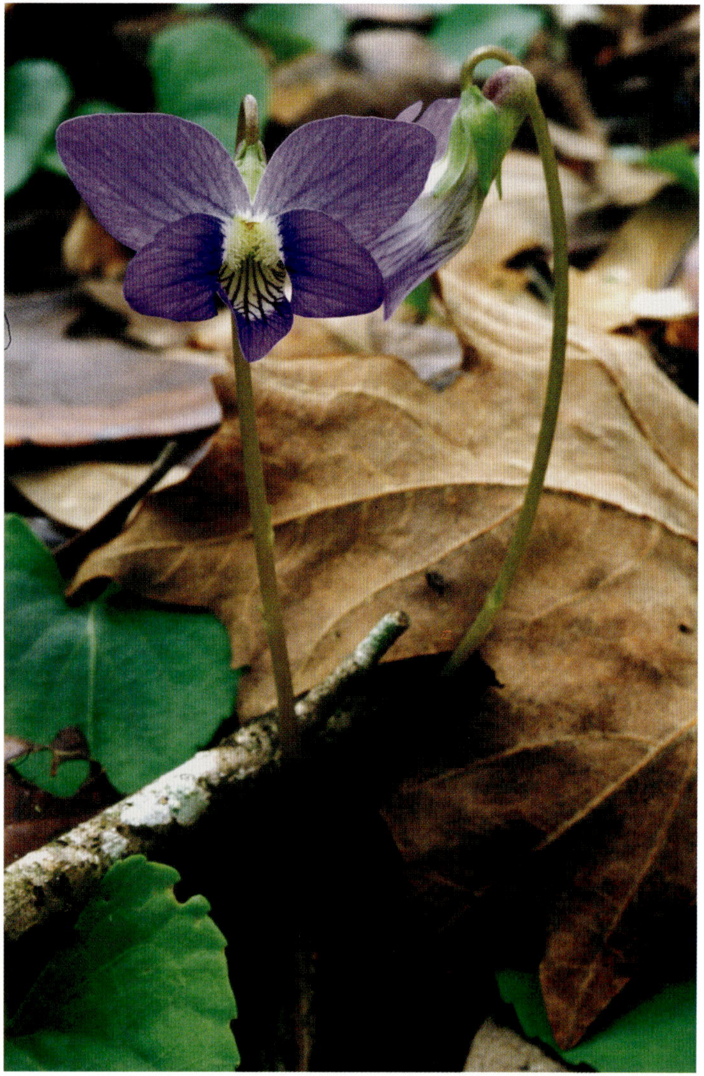

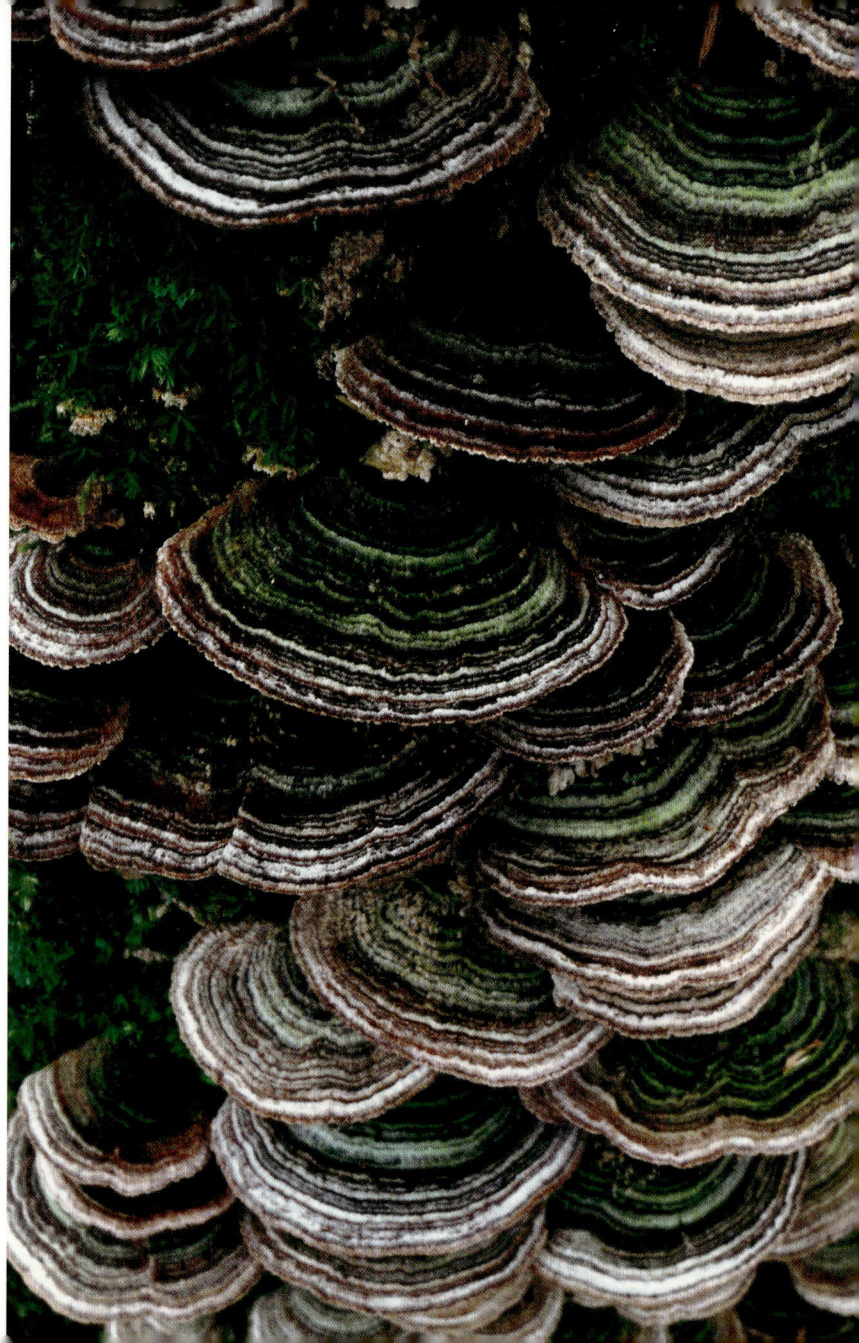

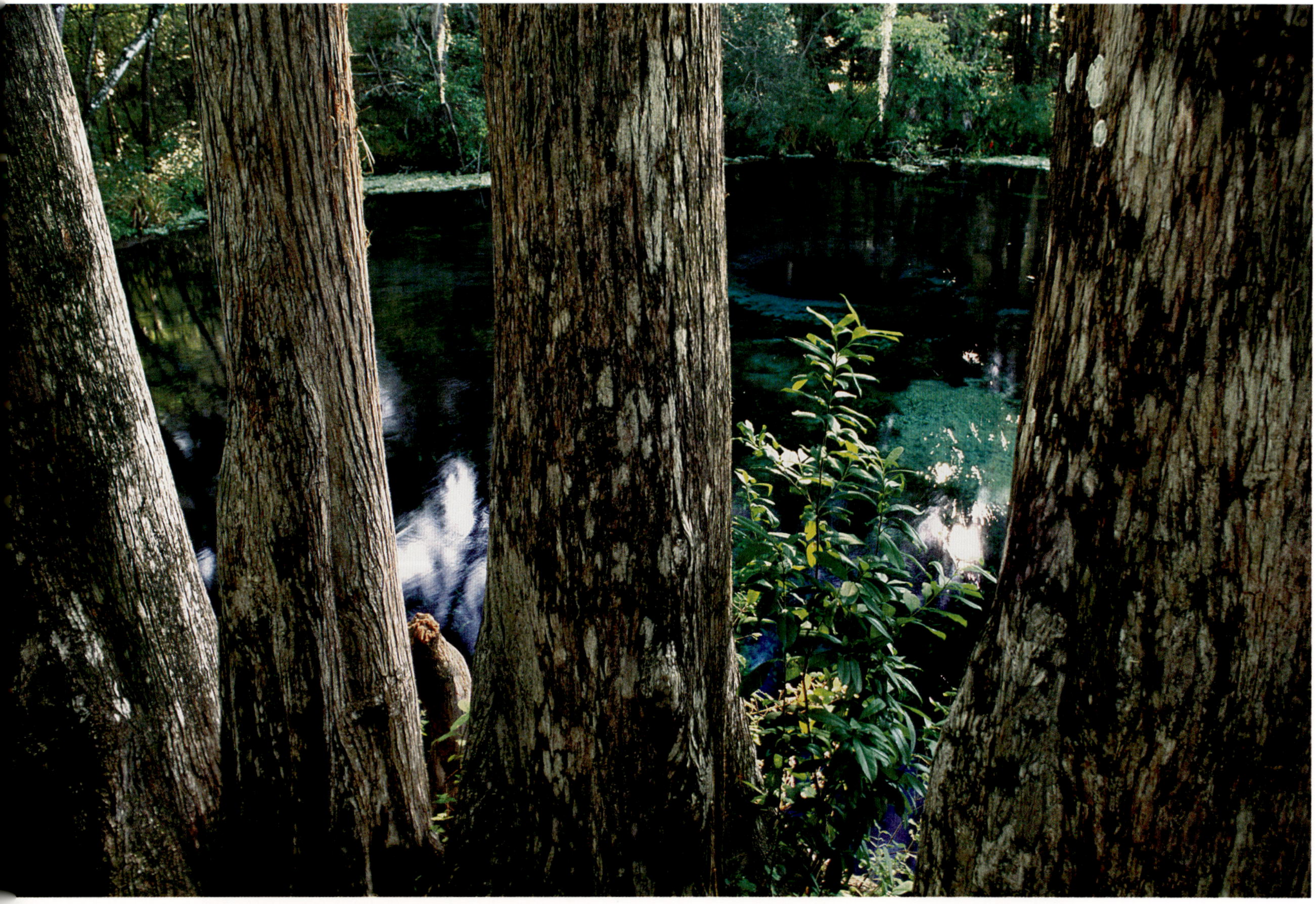

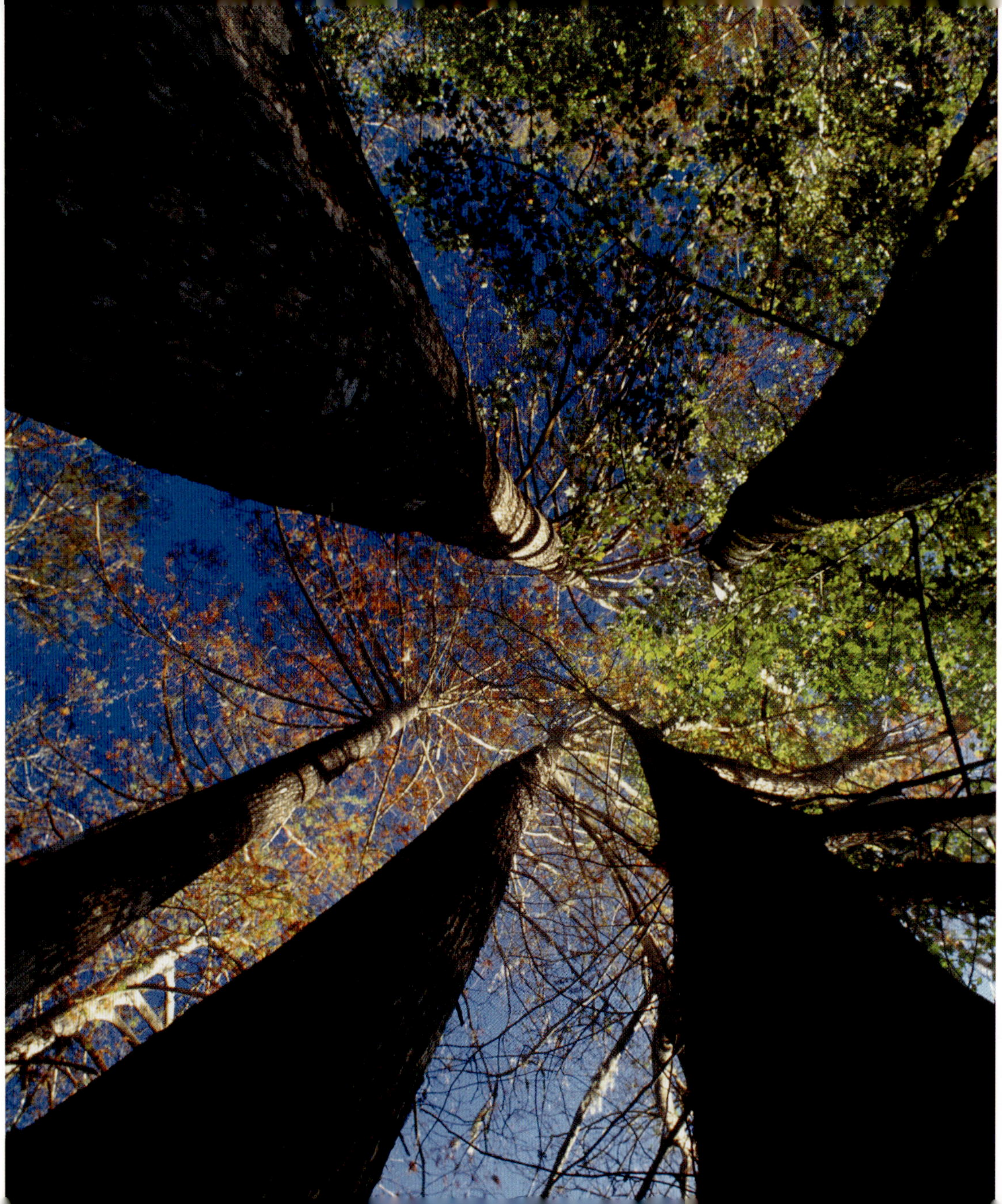

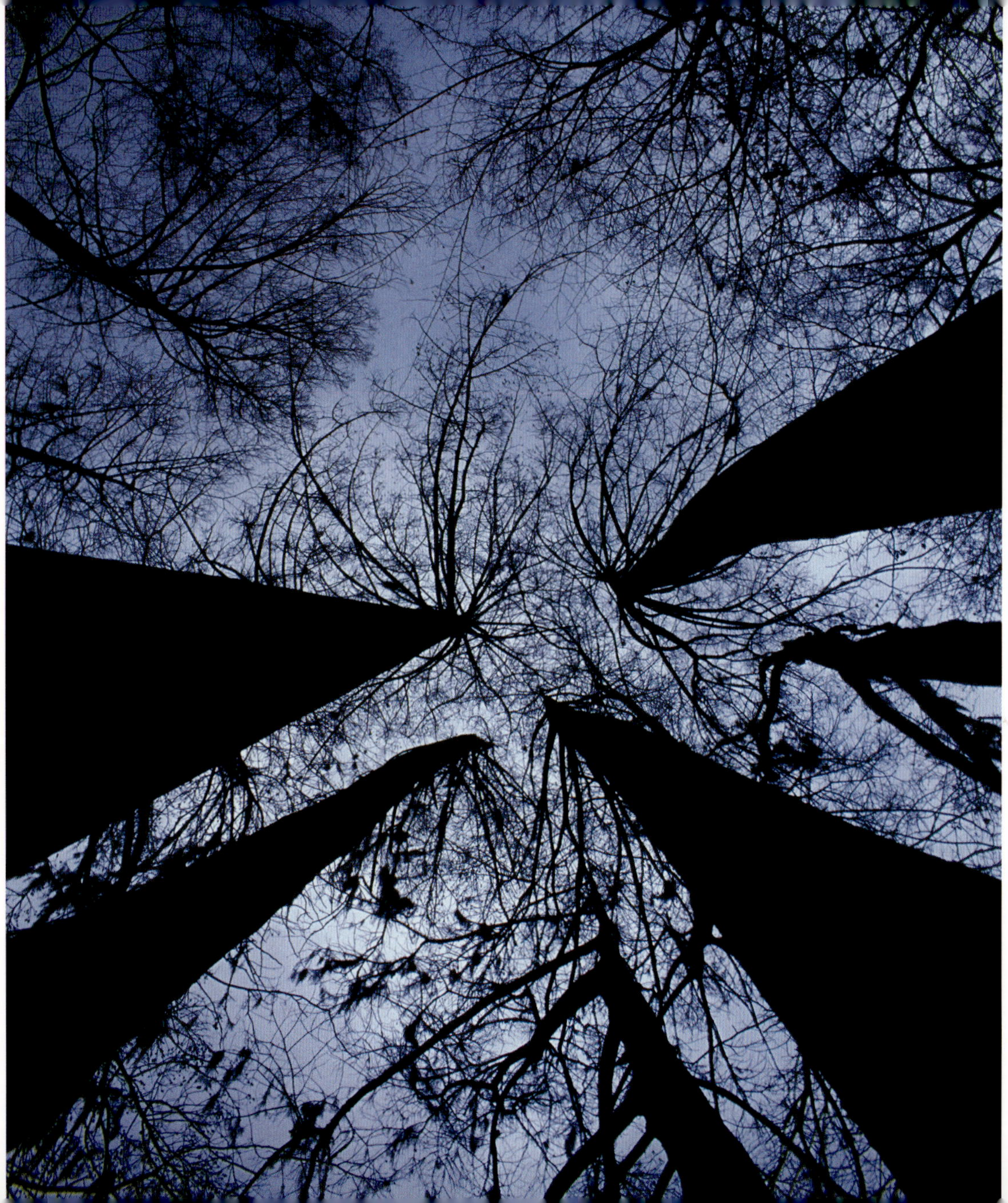

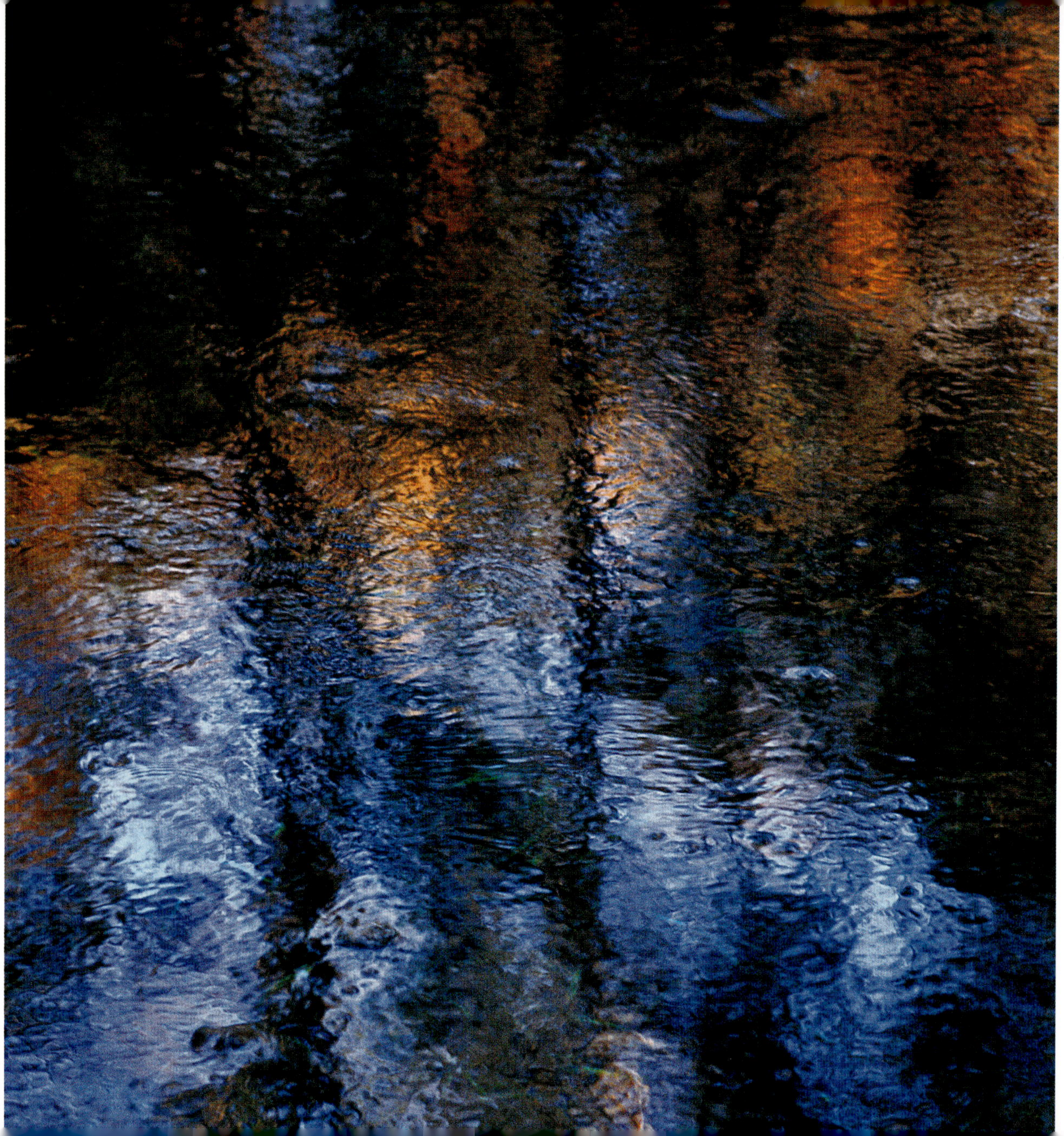

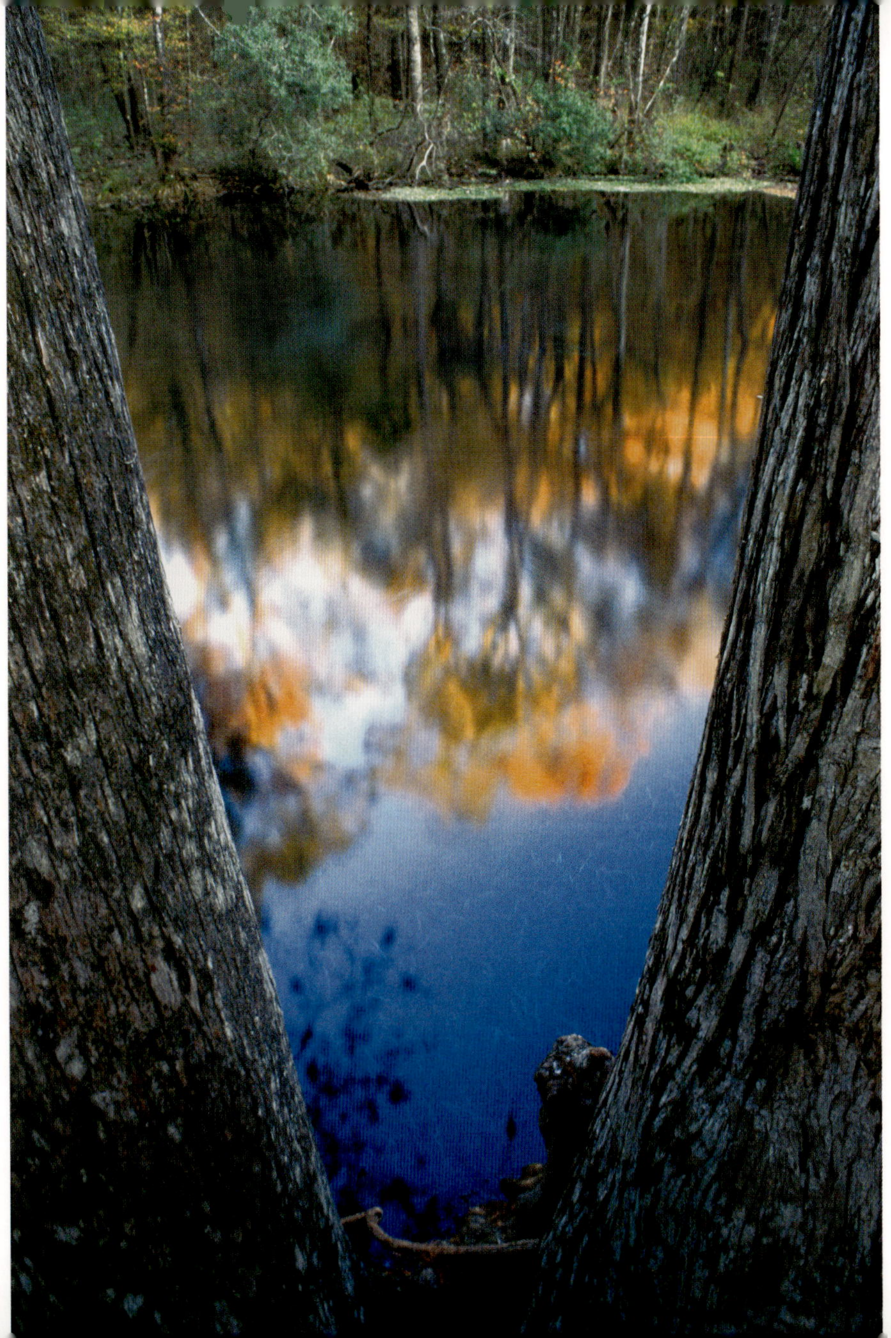

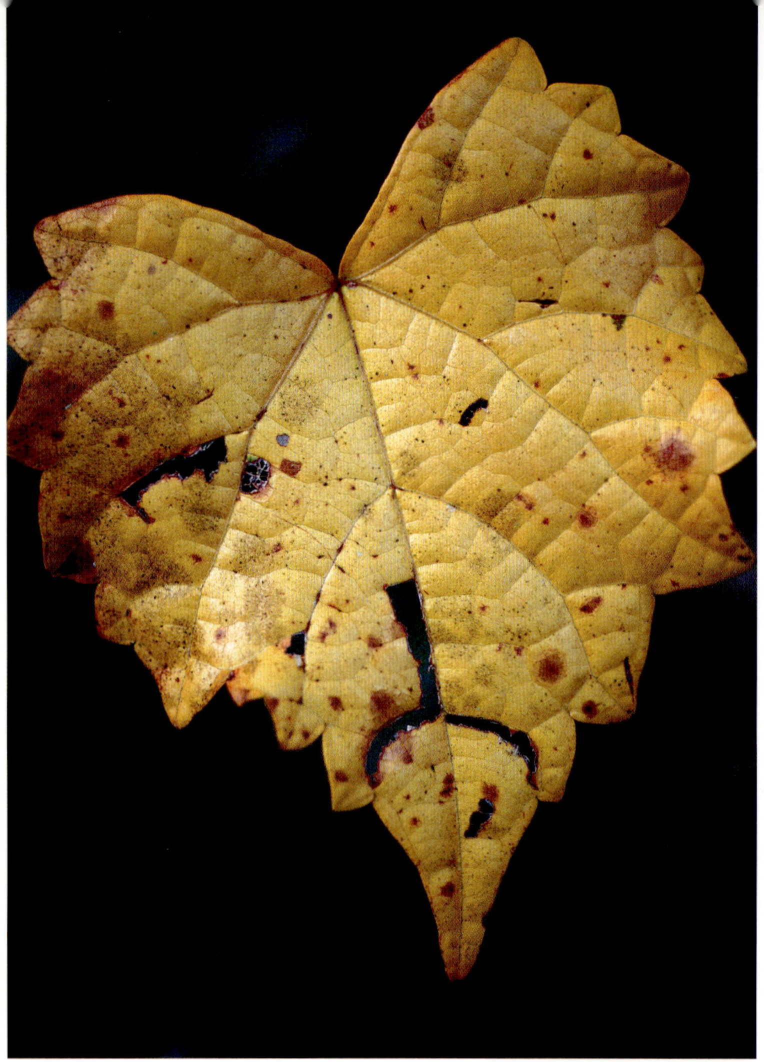

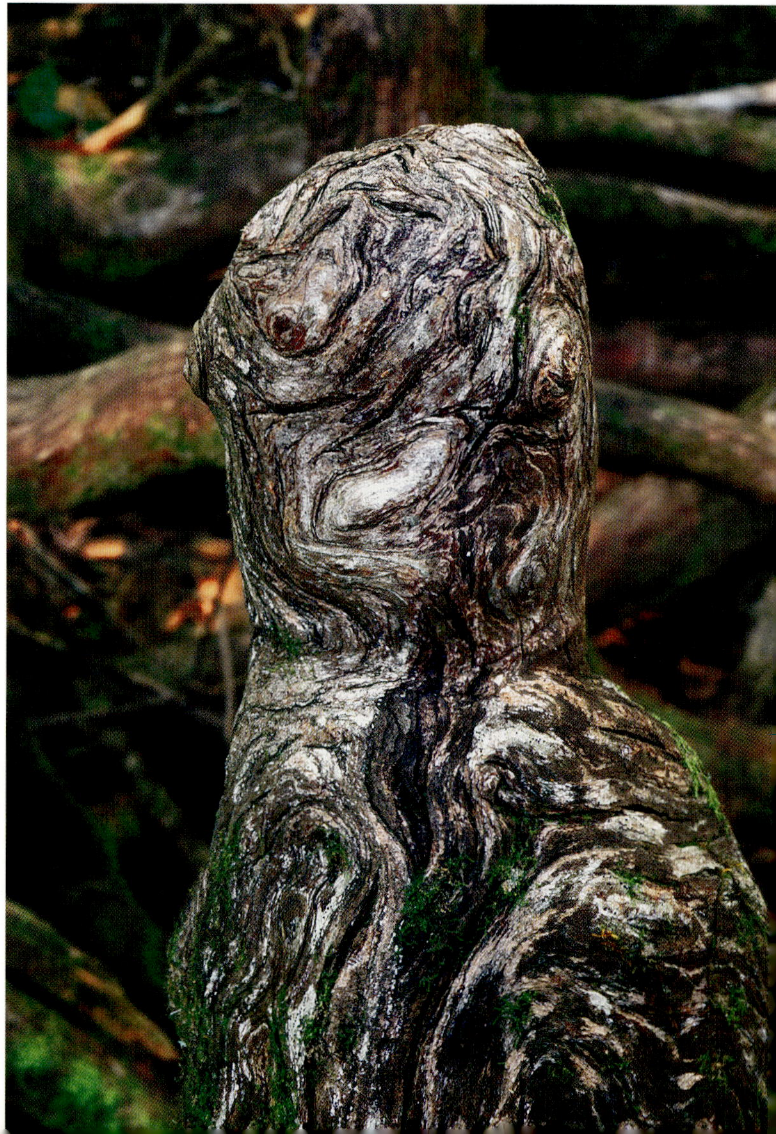

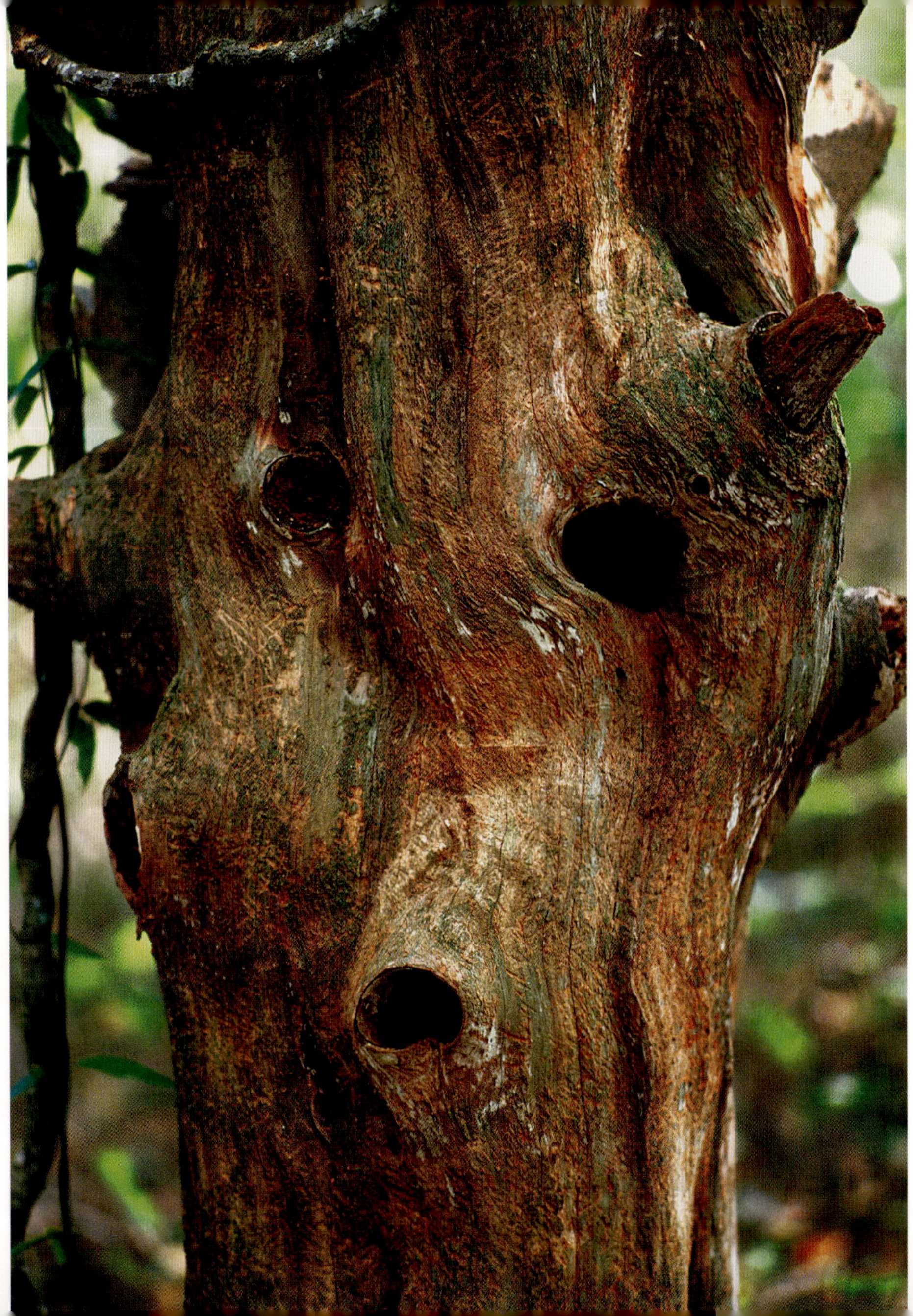

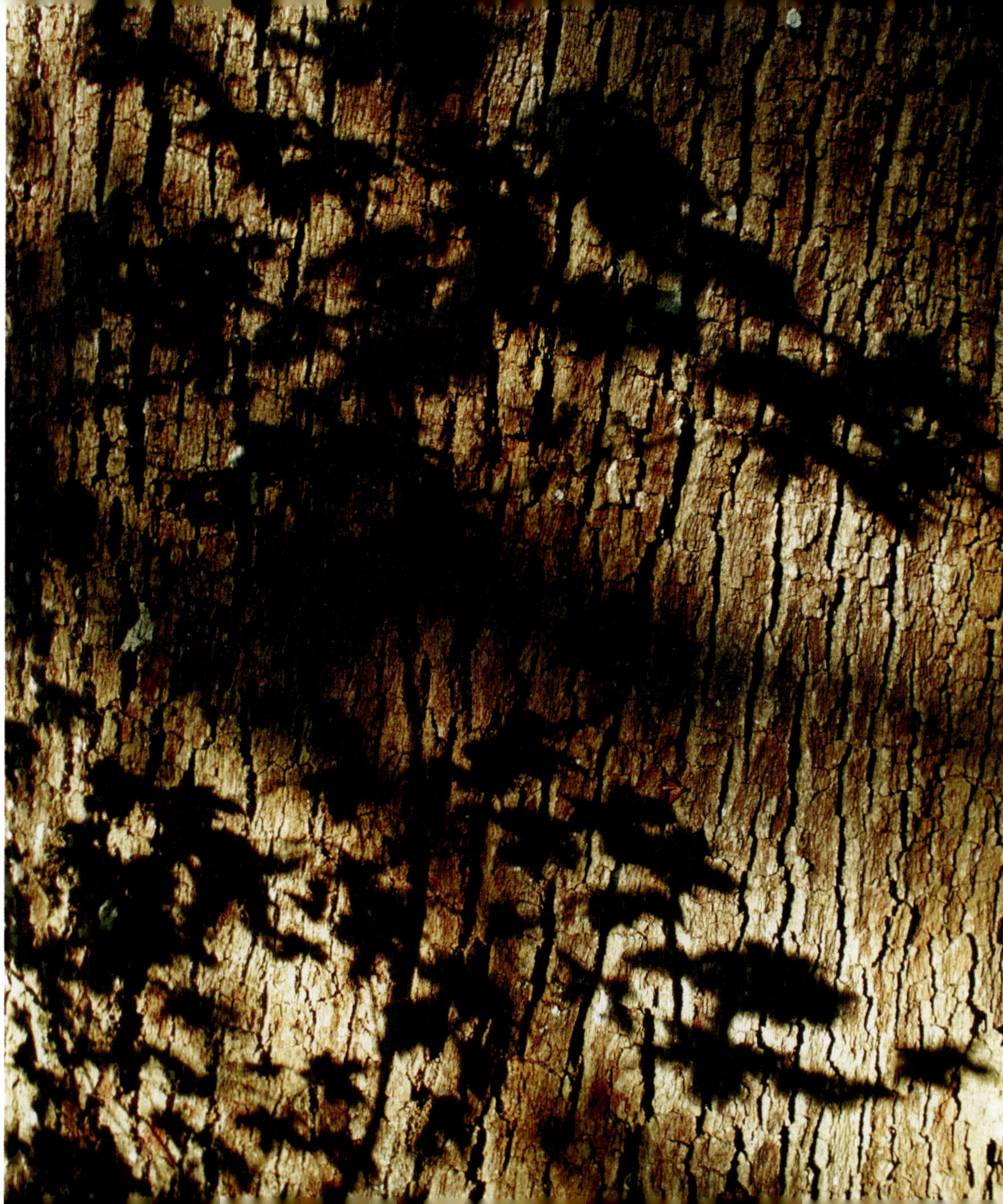

be still—reflect in the shadows—the slow dance of fading light—and the gifts received of the day . . .

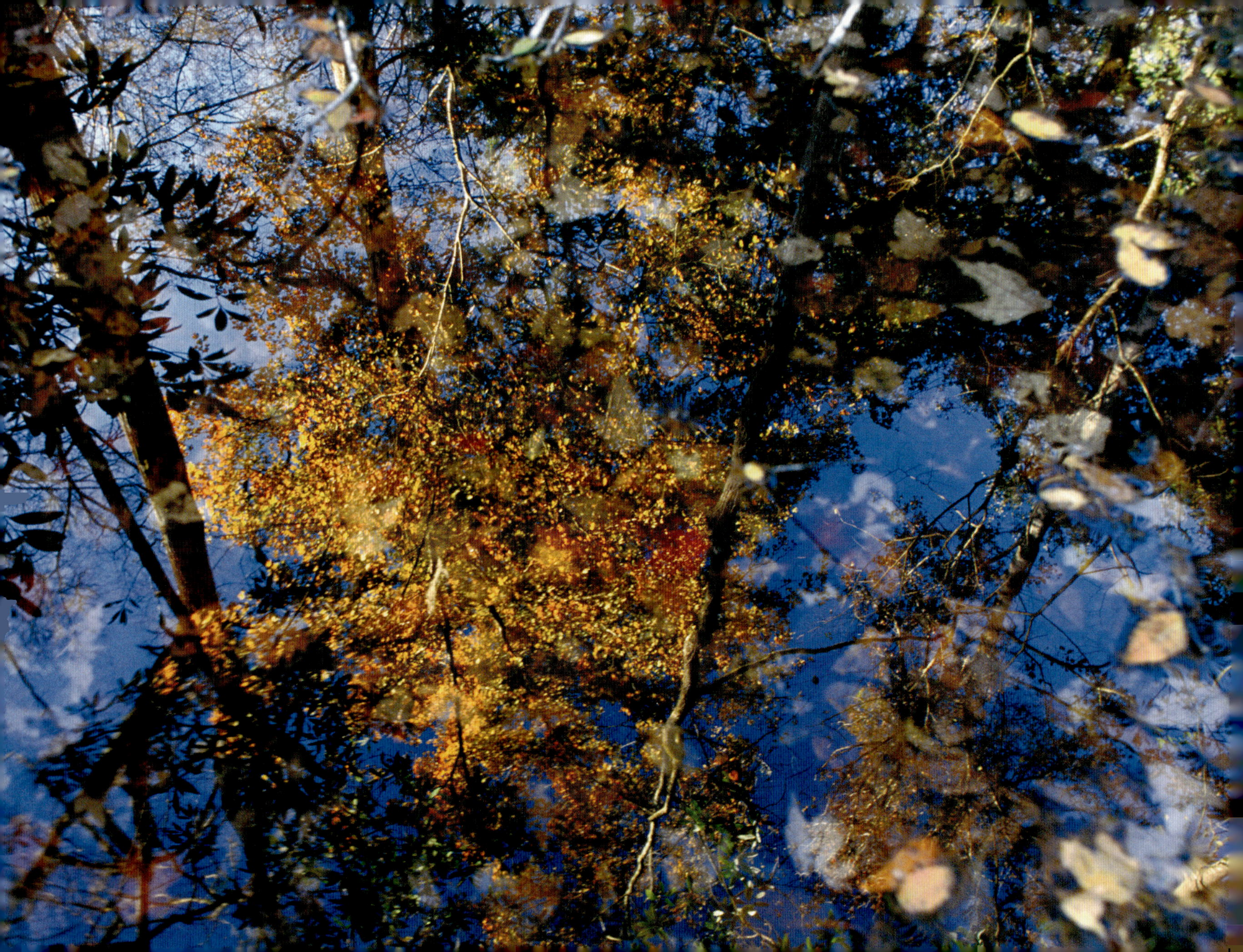

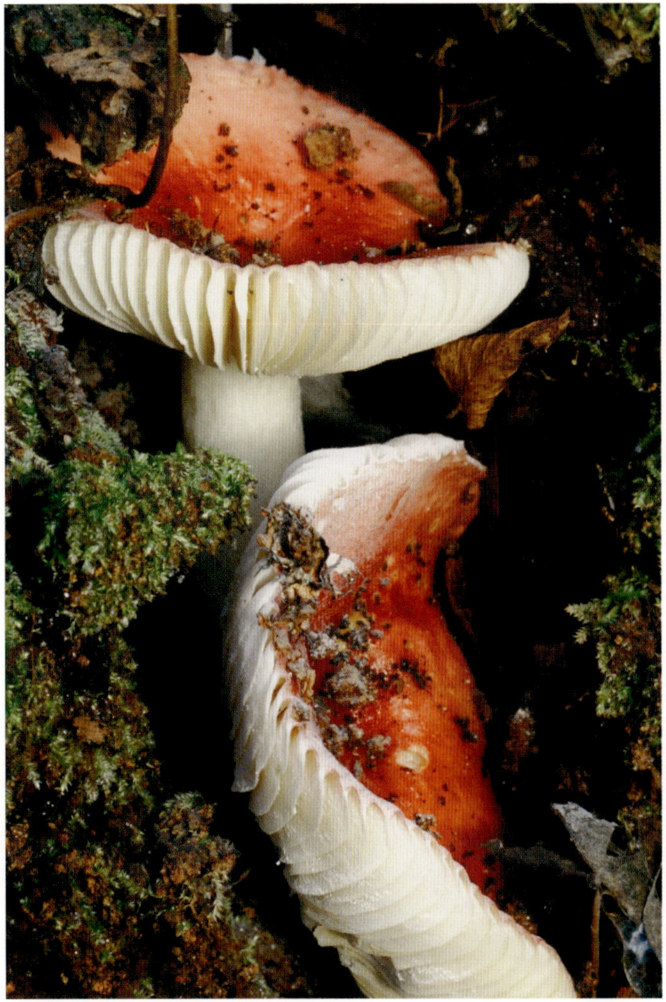

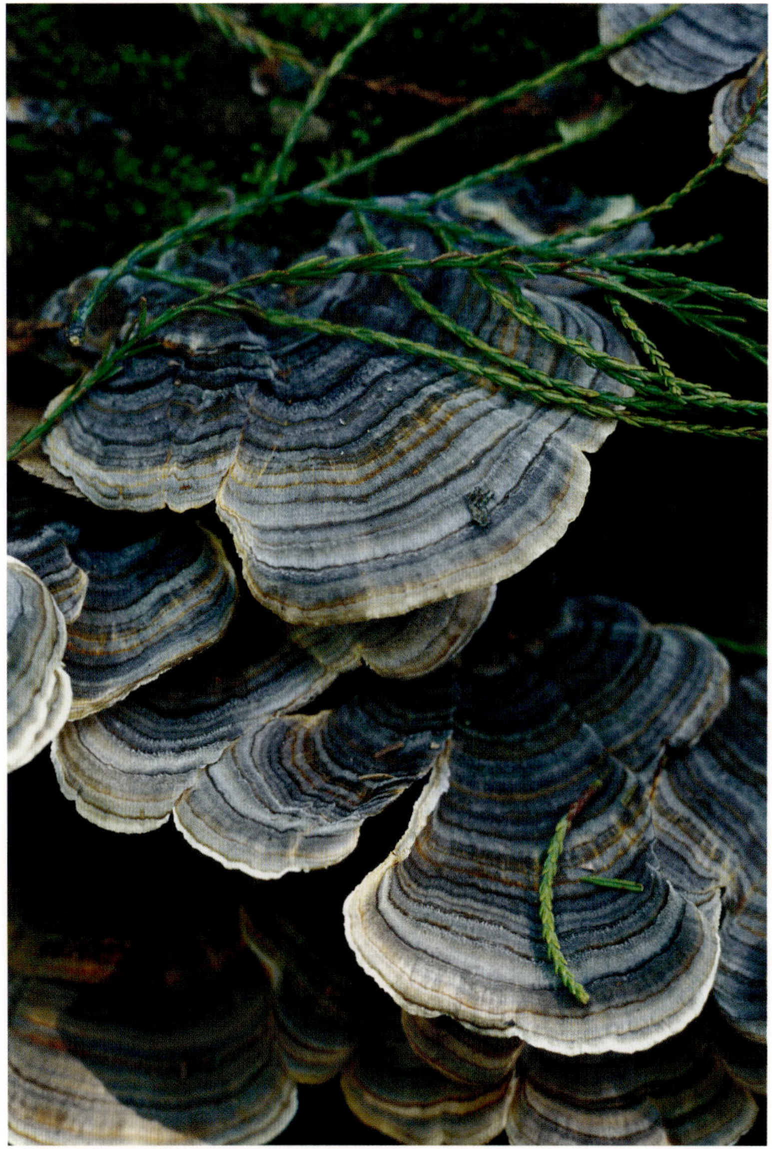

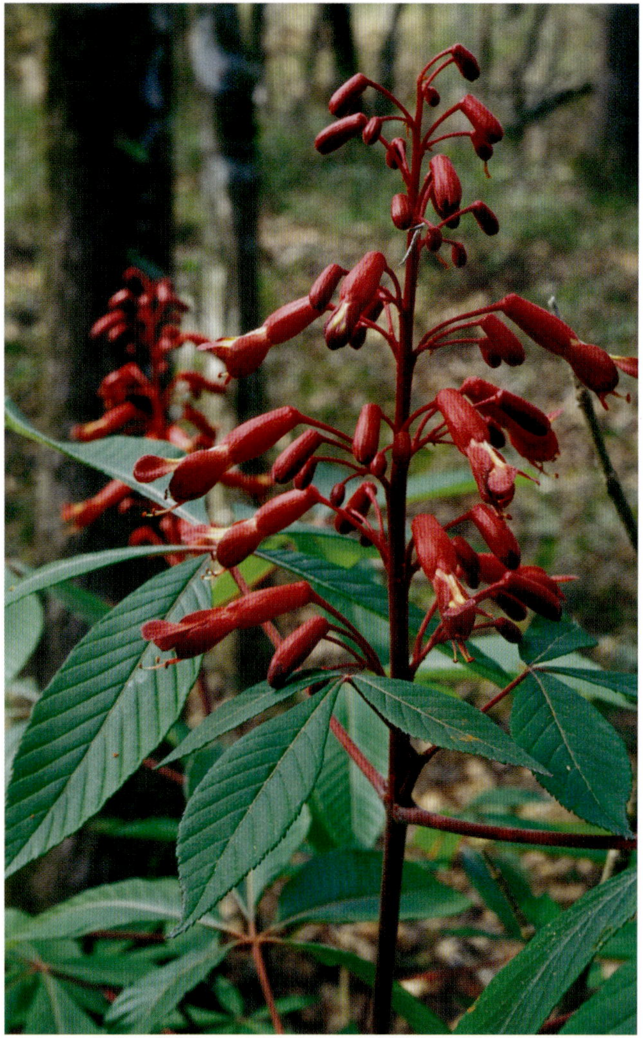

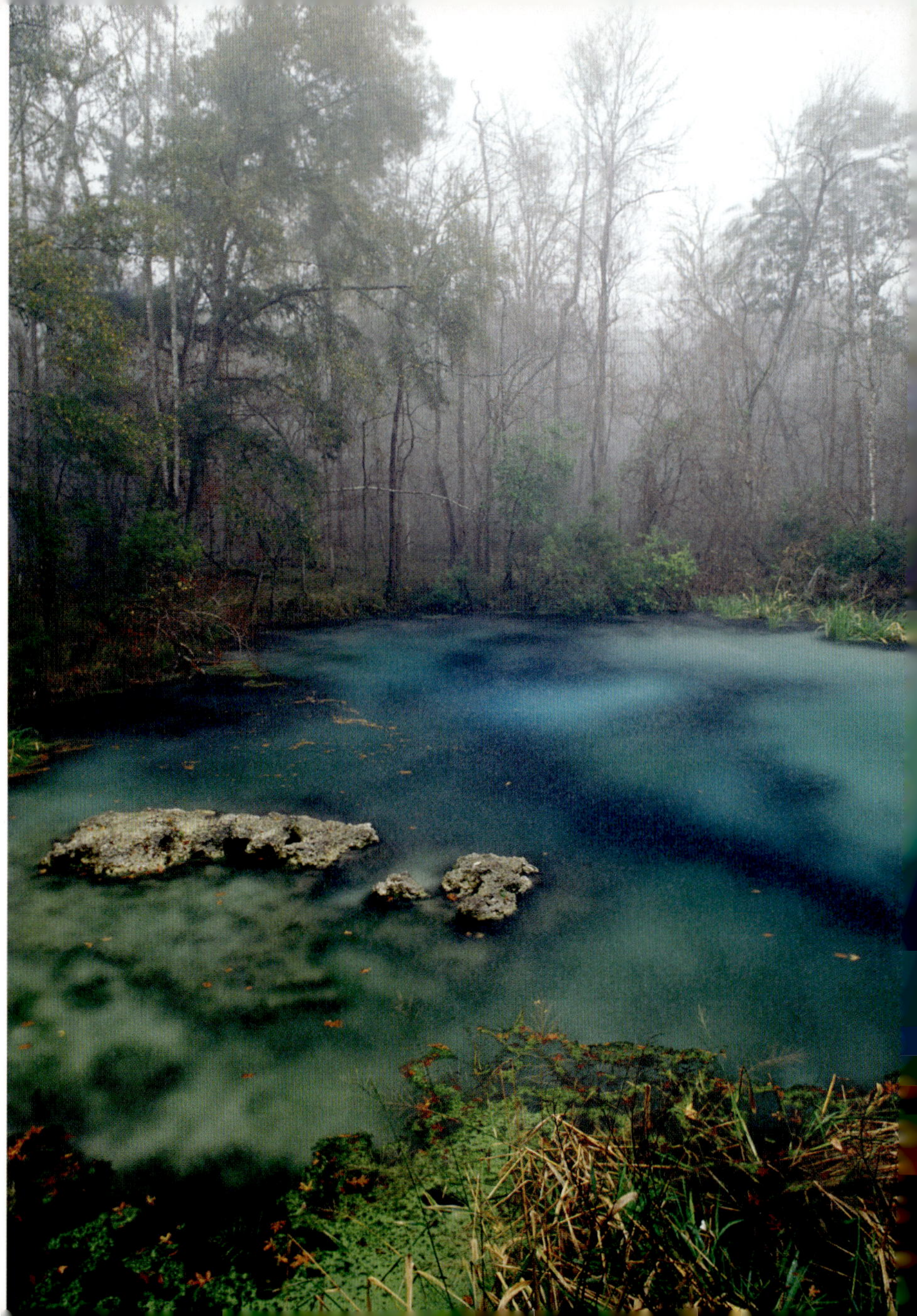

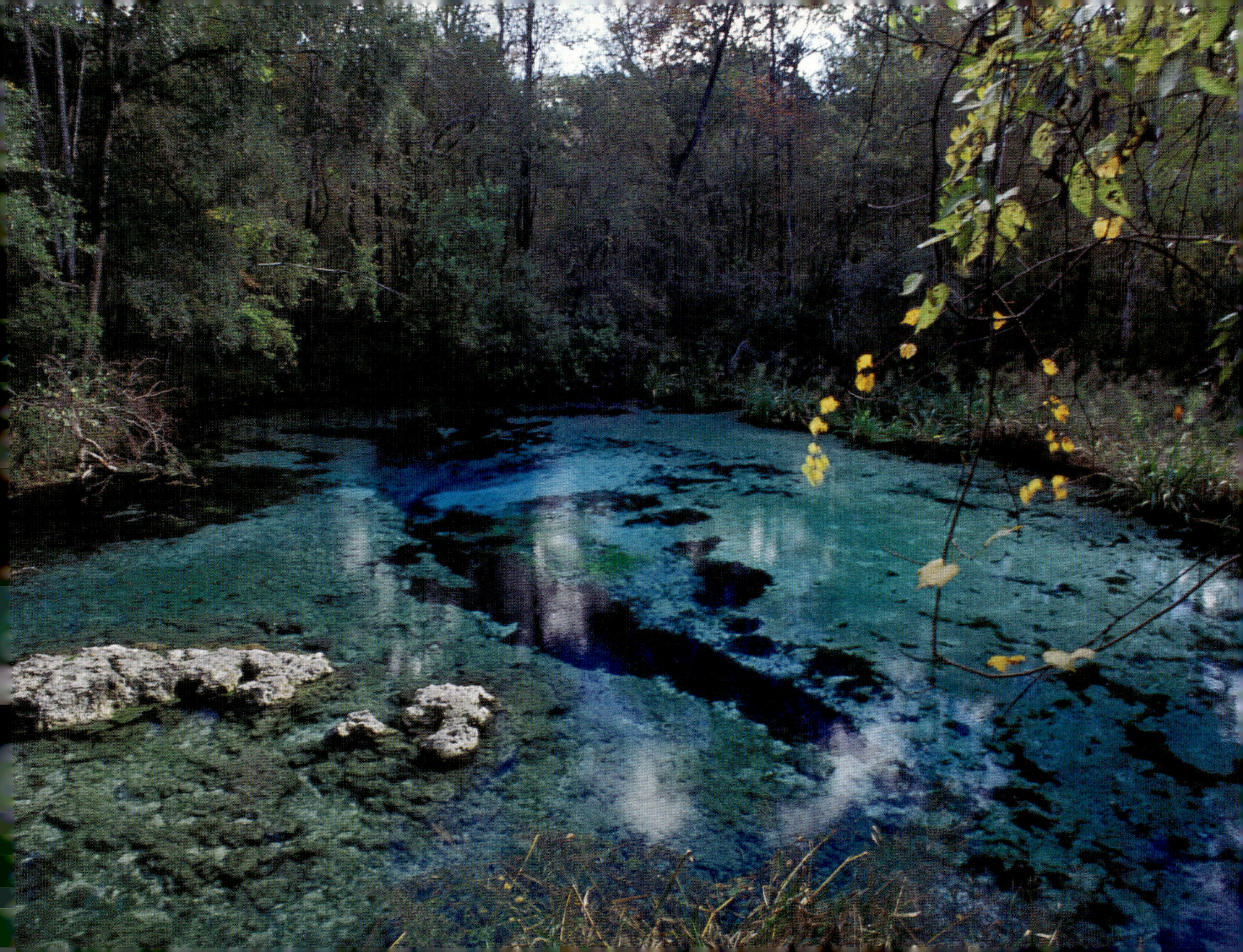

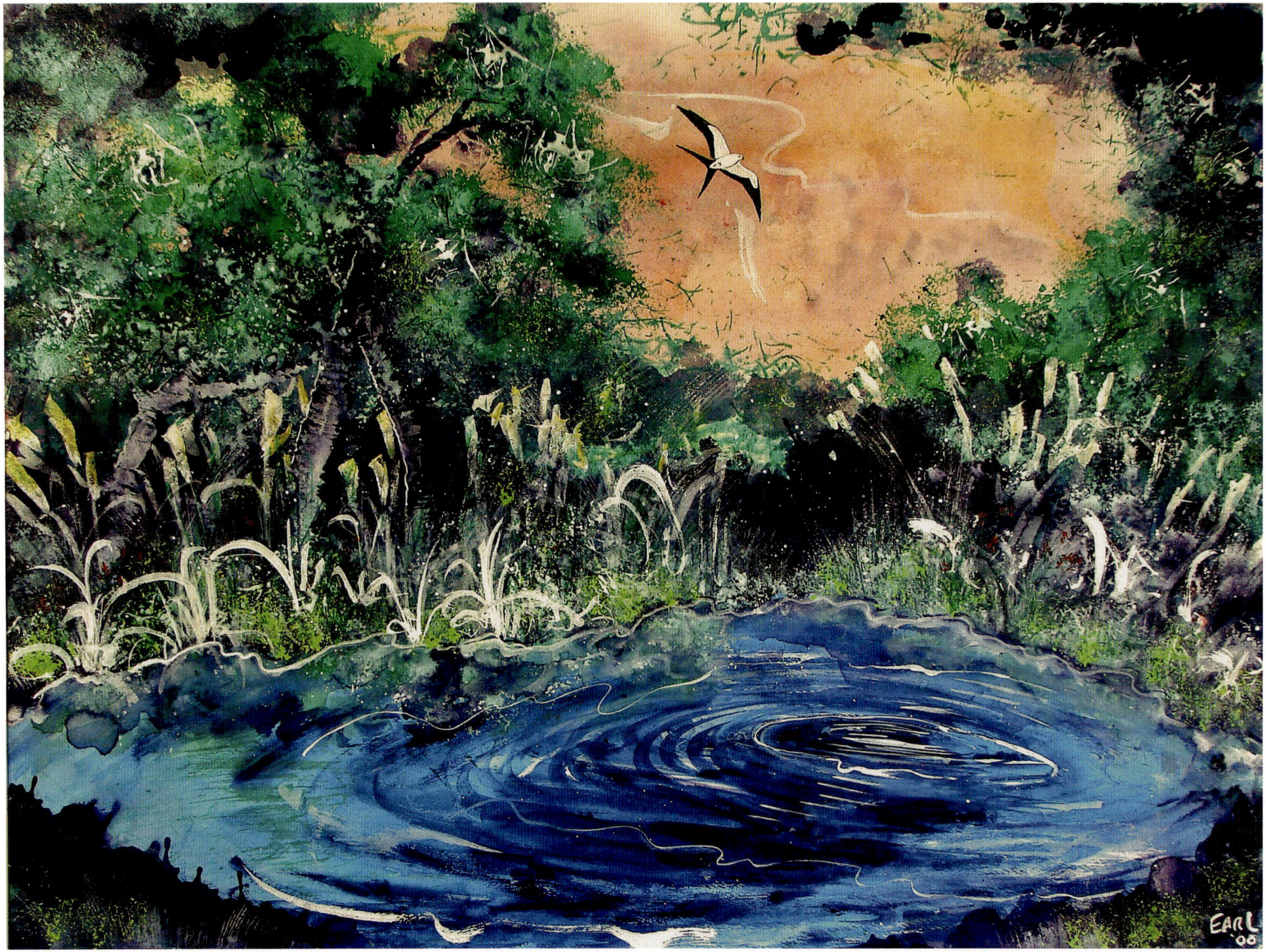
EARL
'00

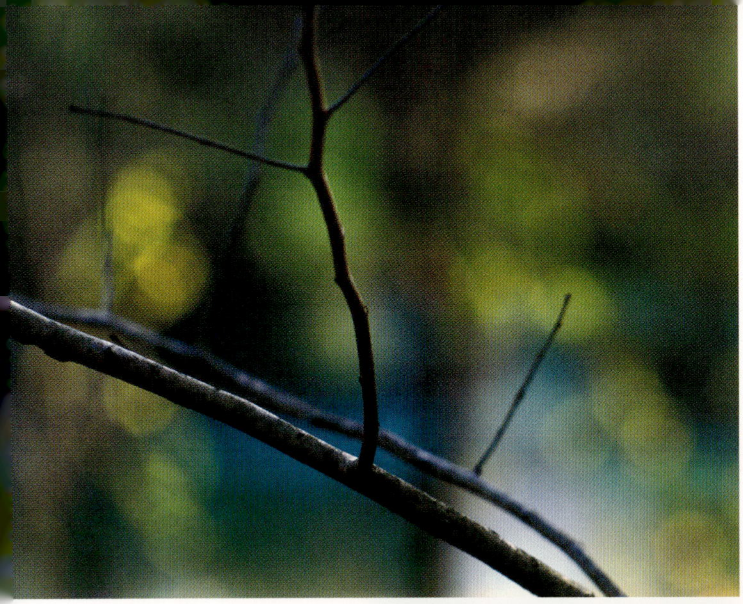

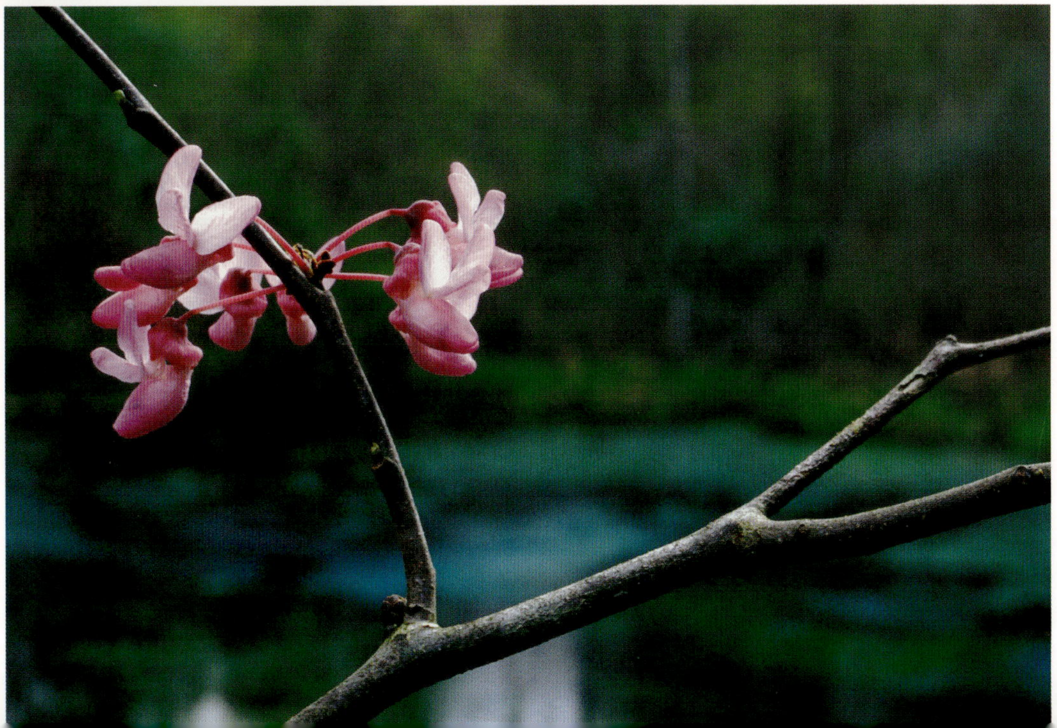

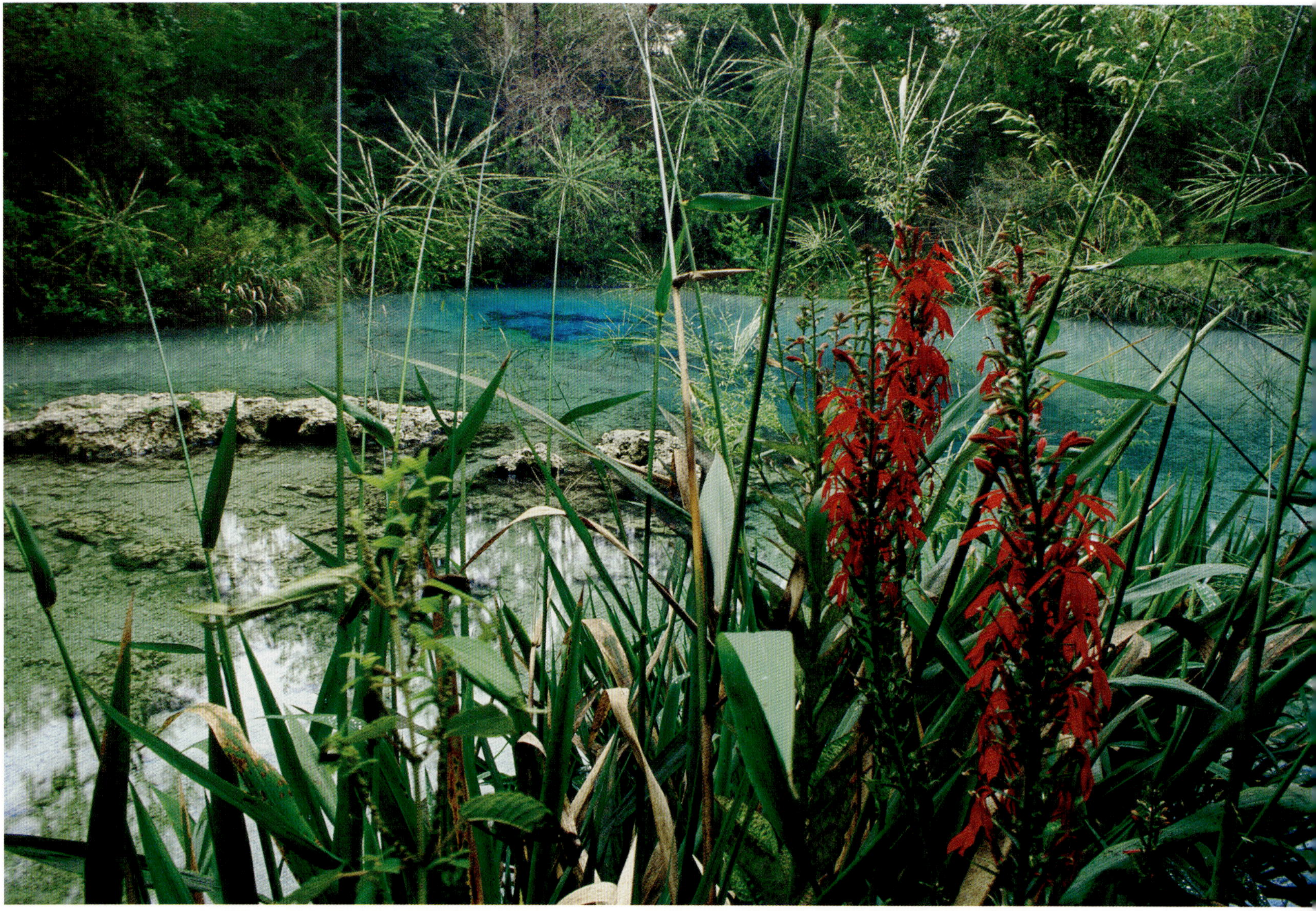

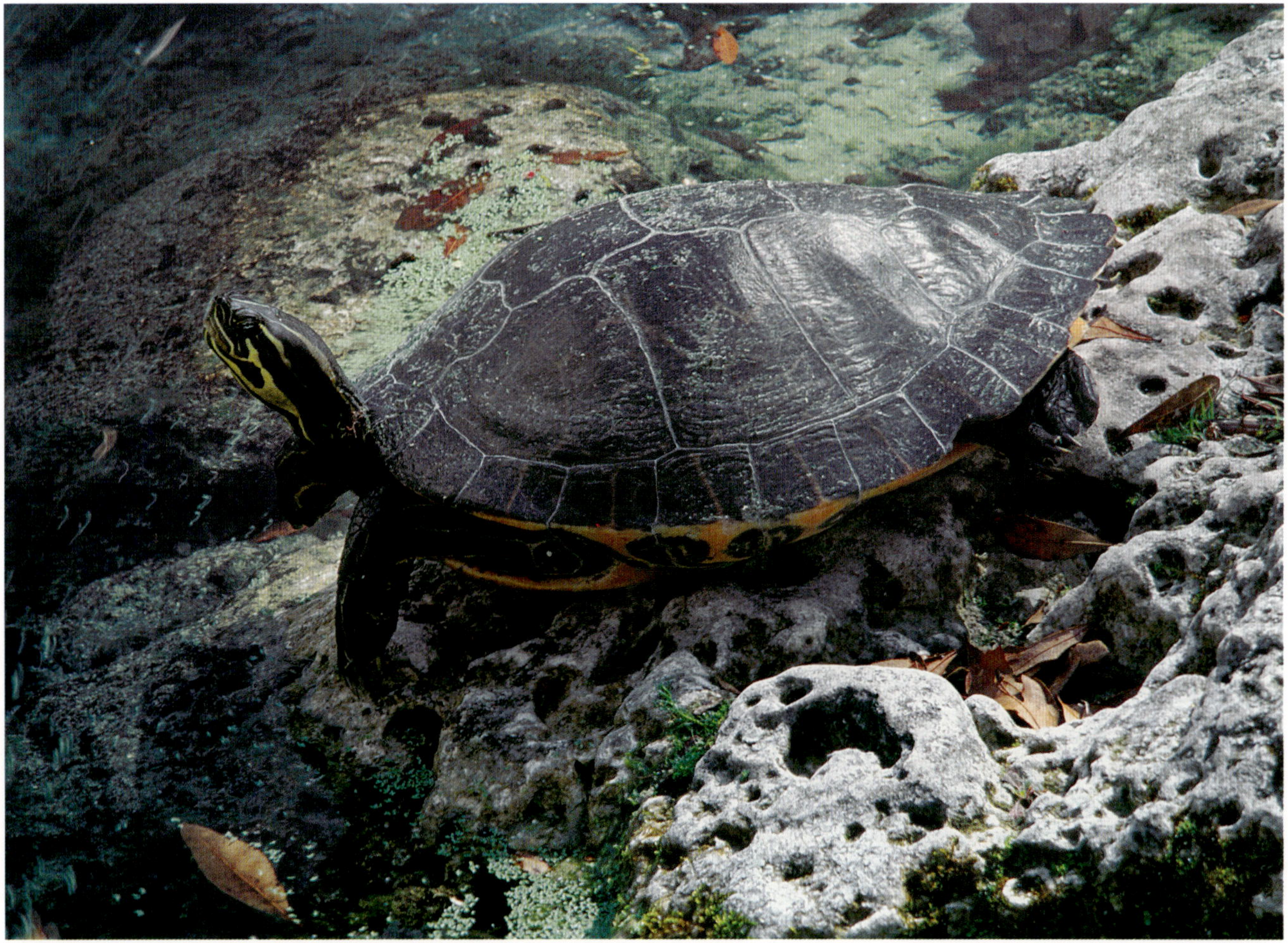

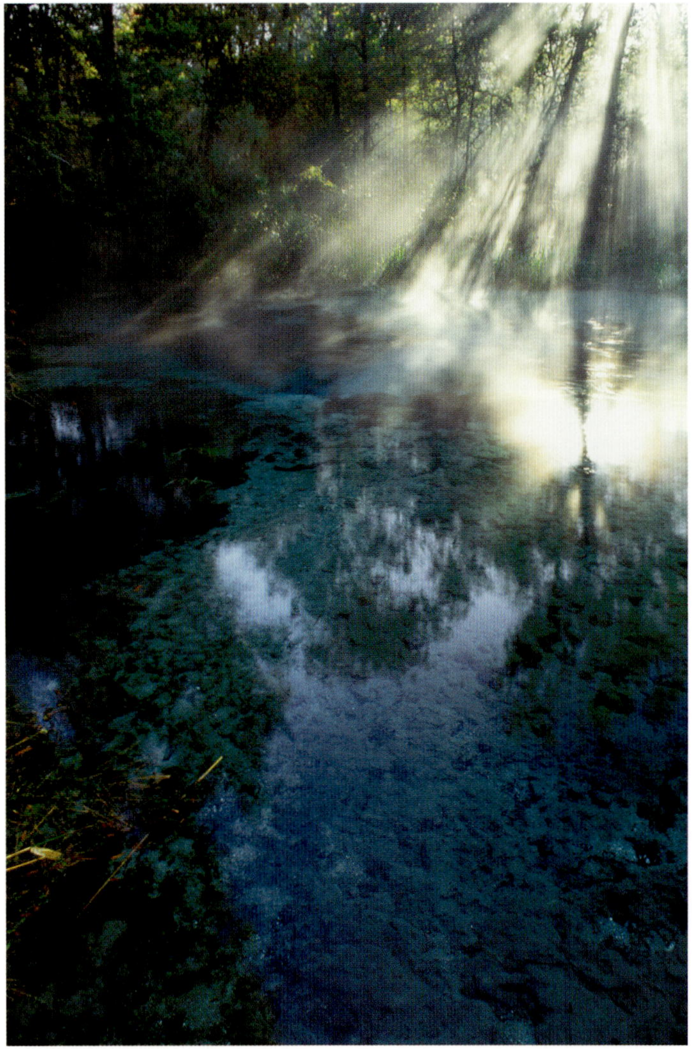

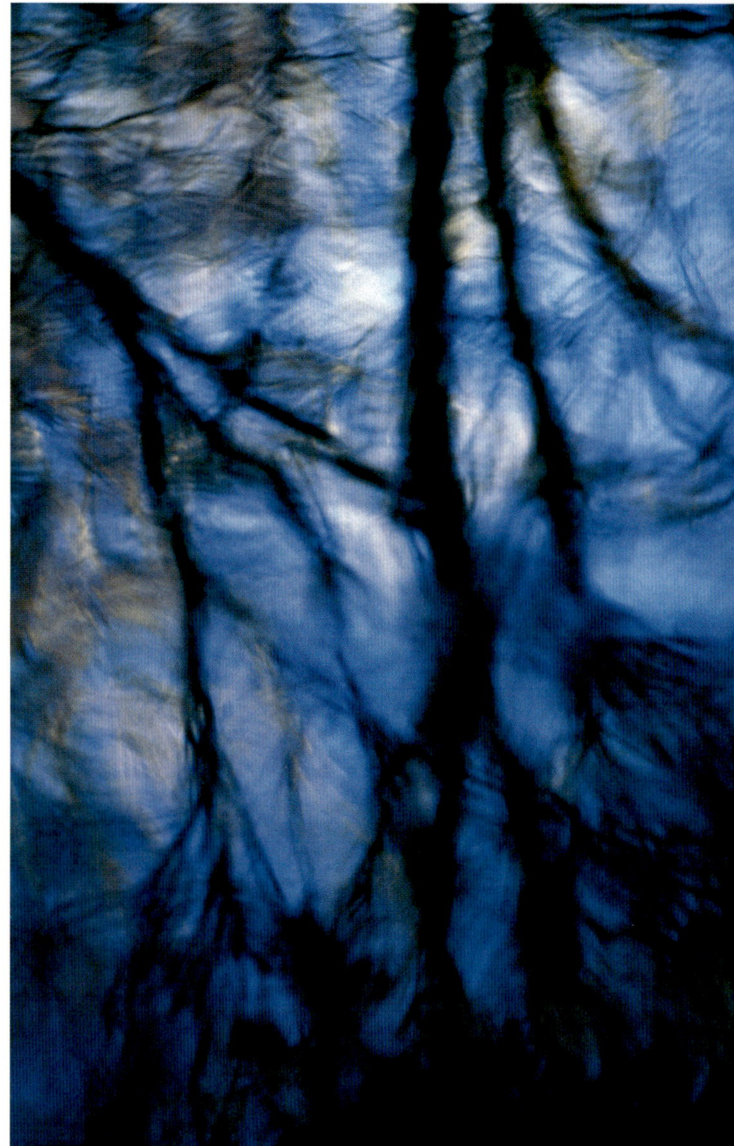

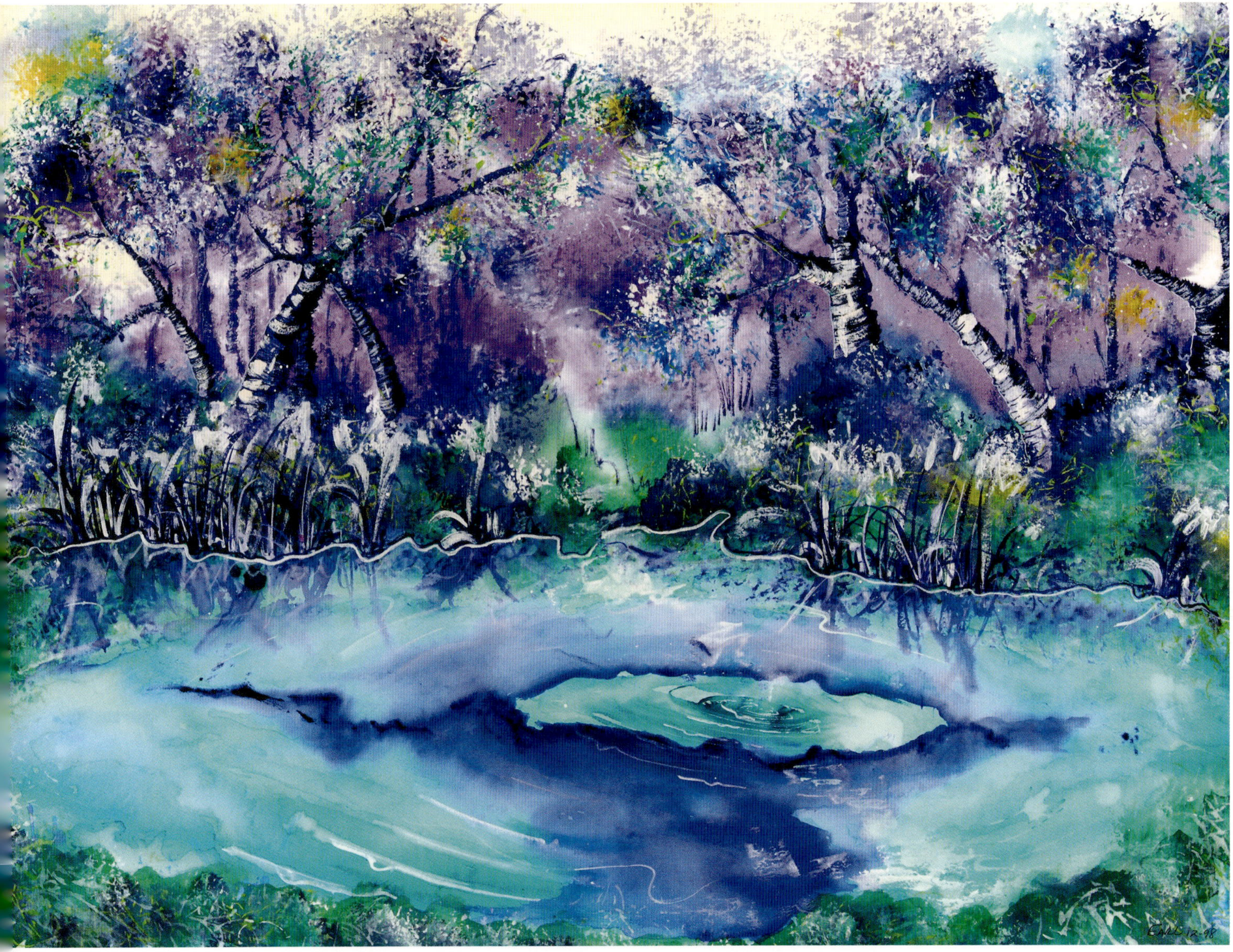

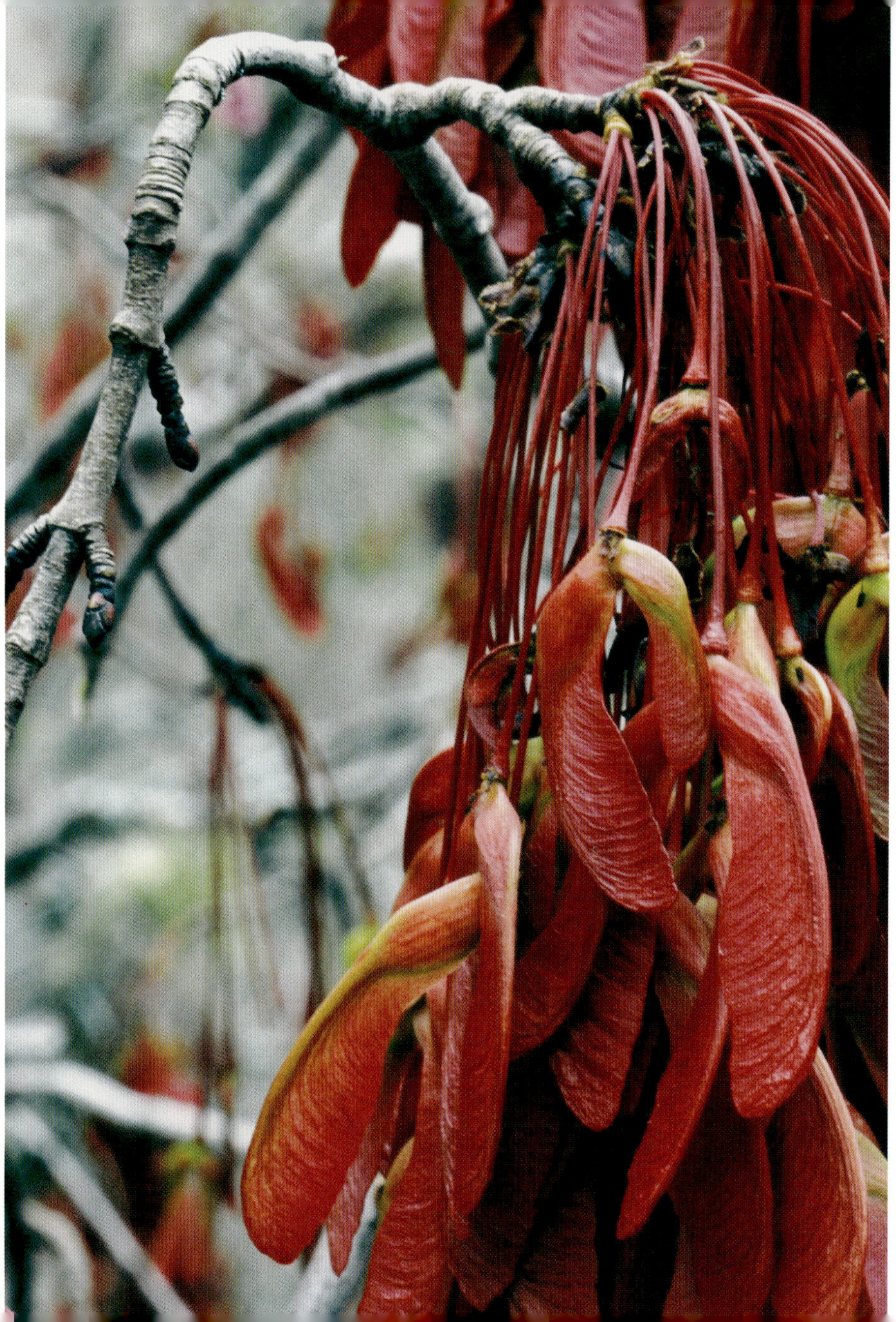

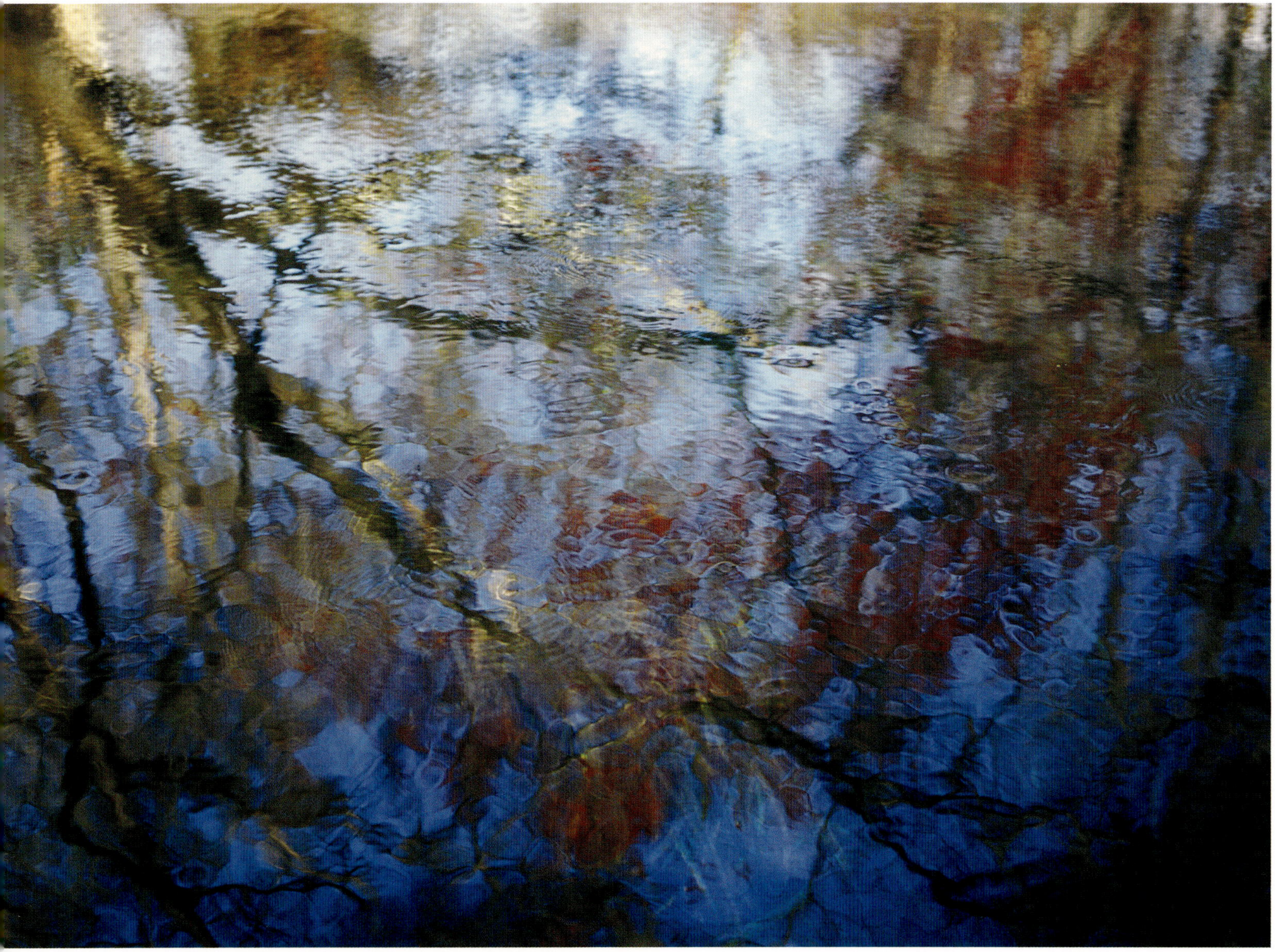

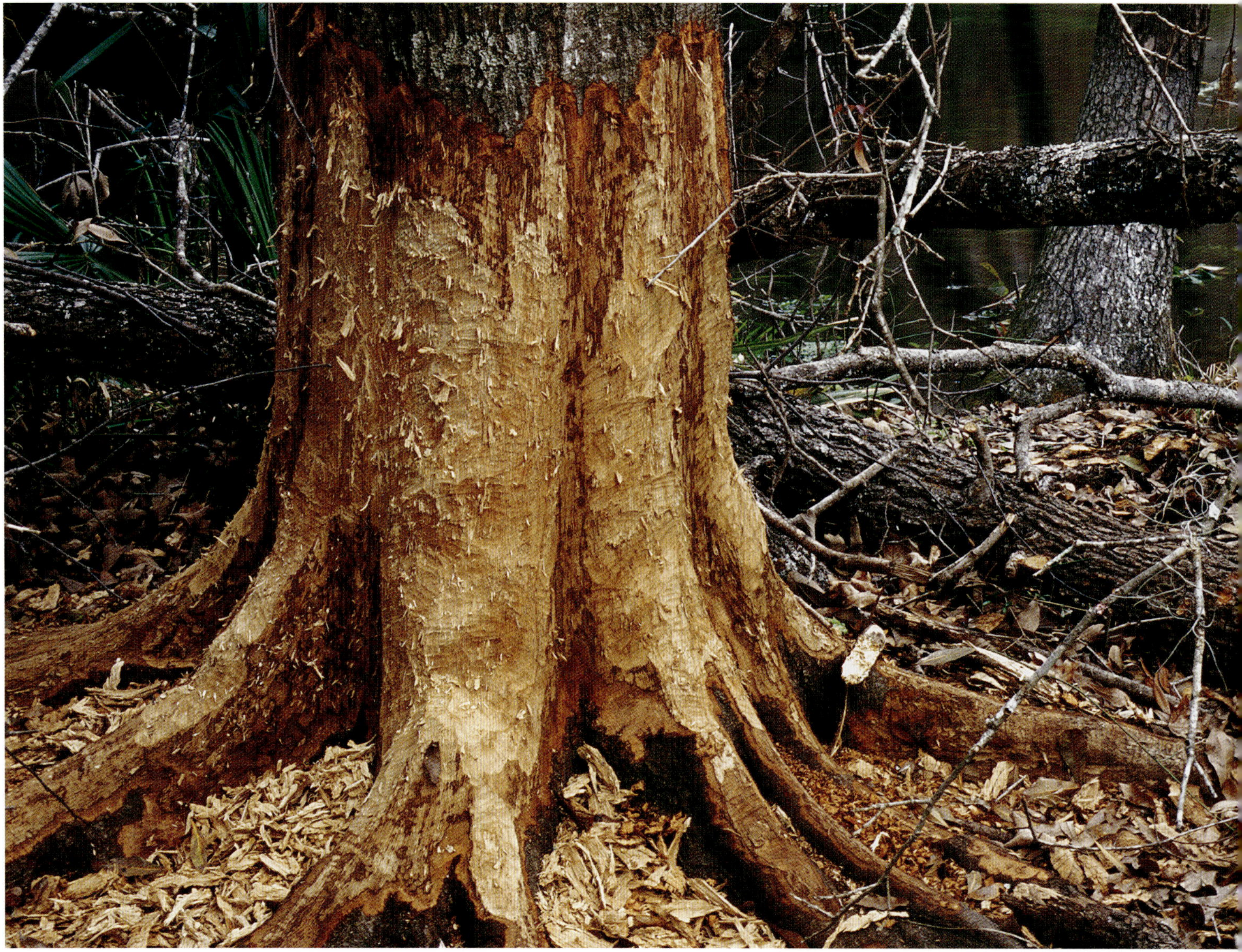

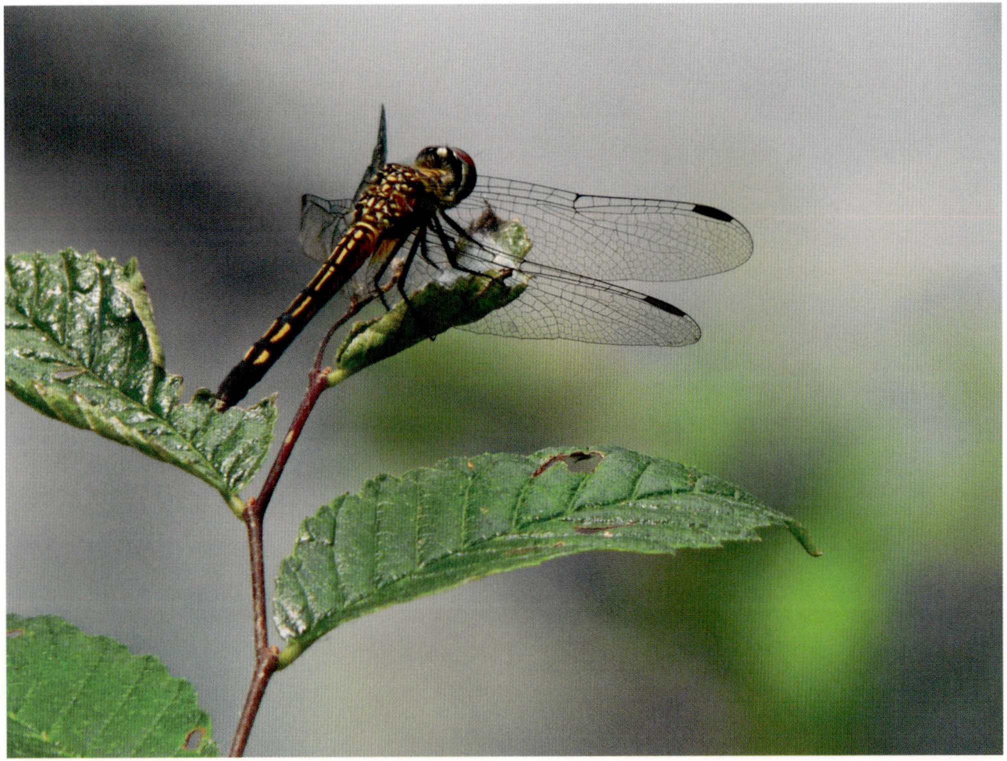

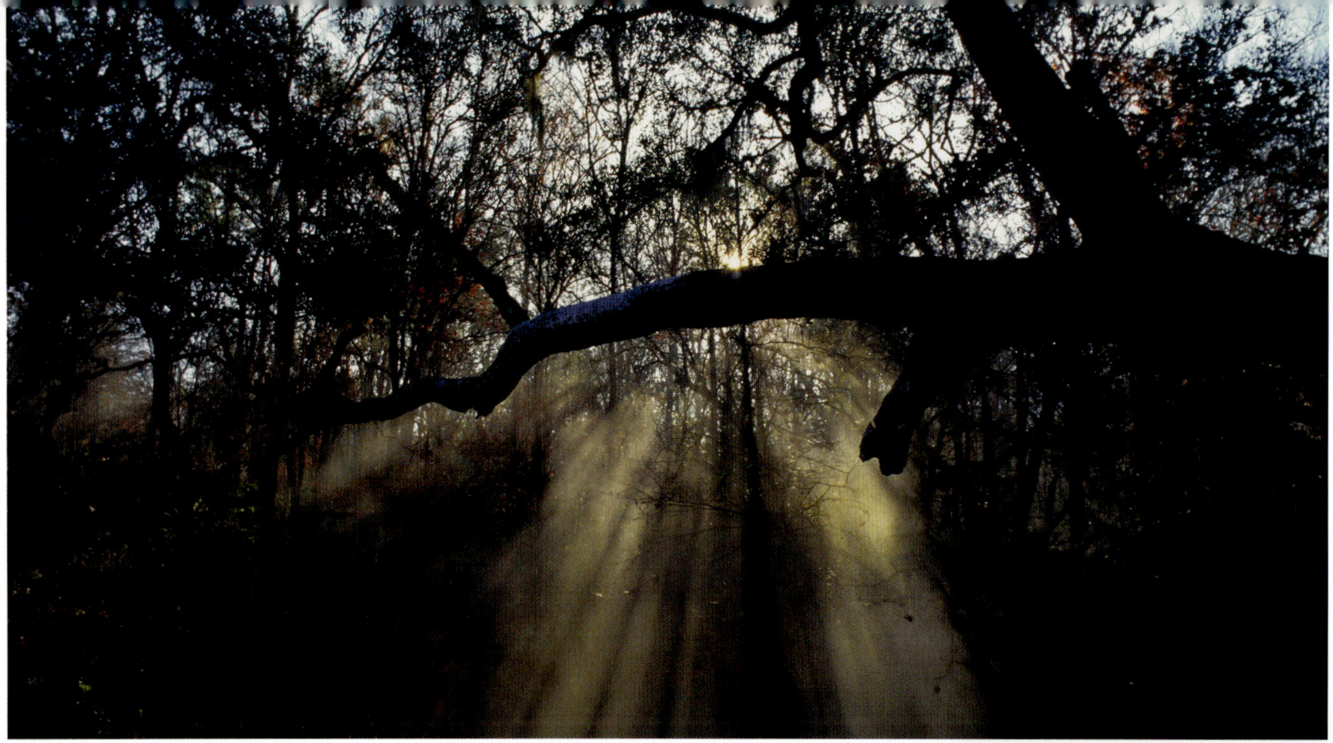

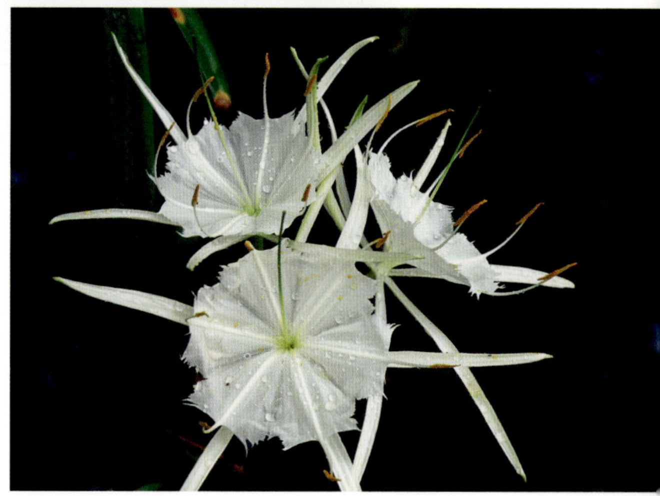

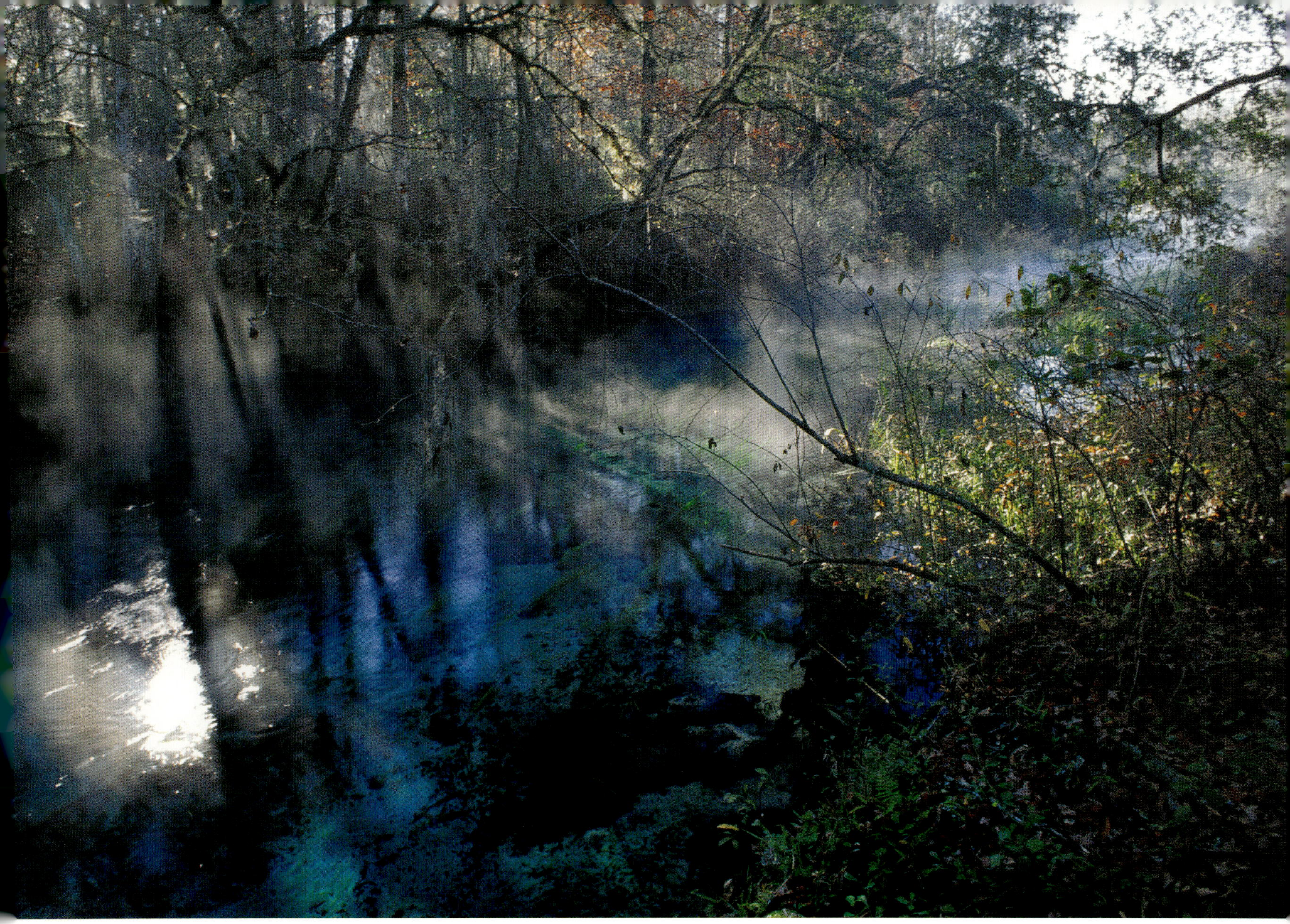

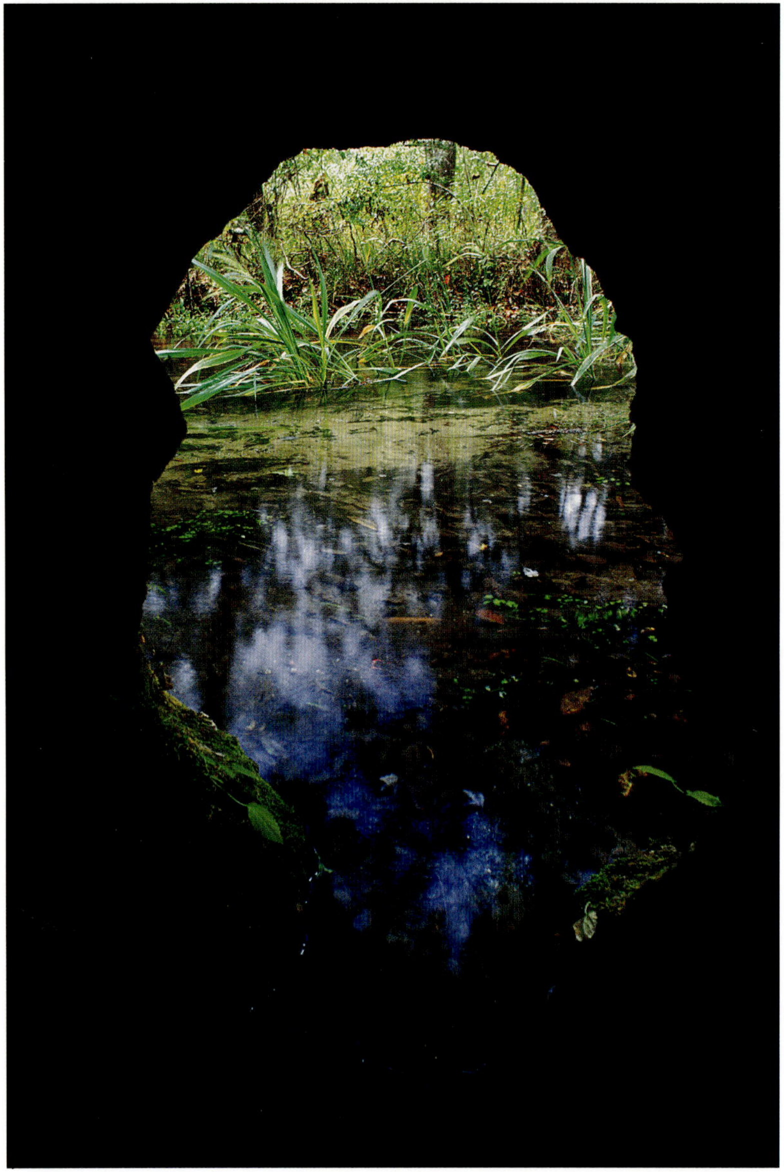
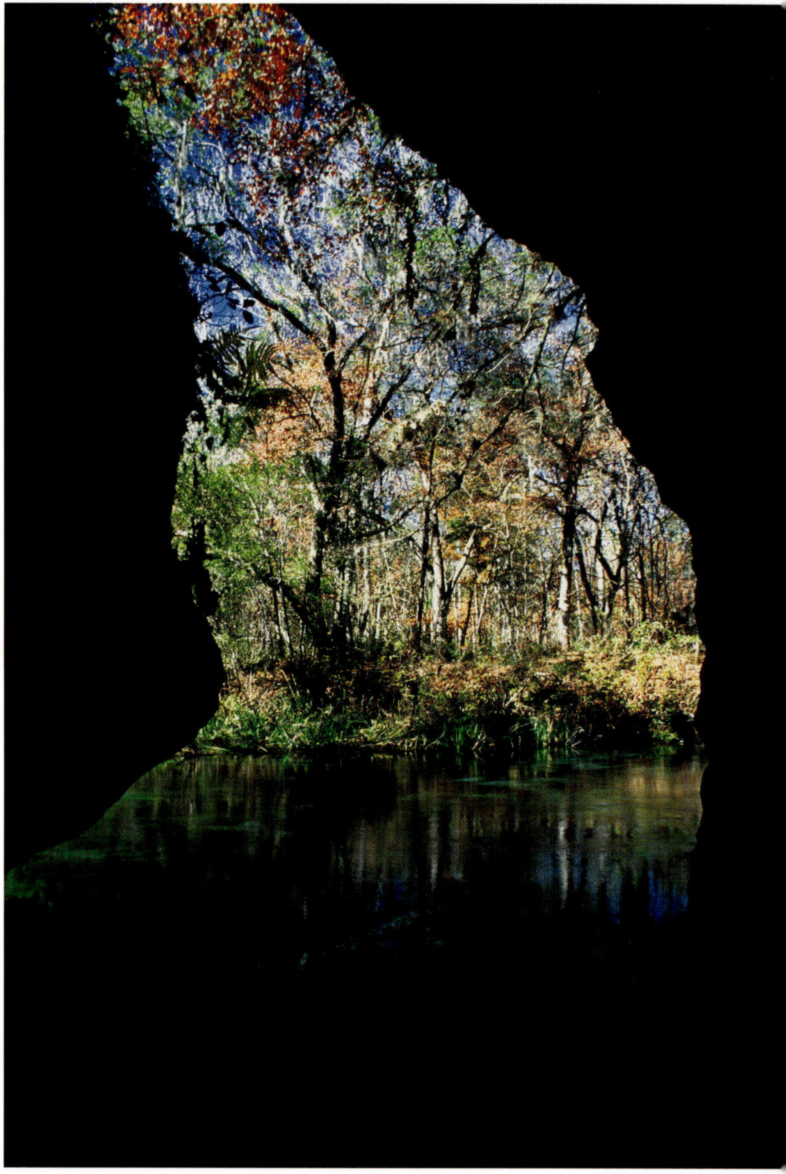

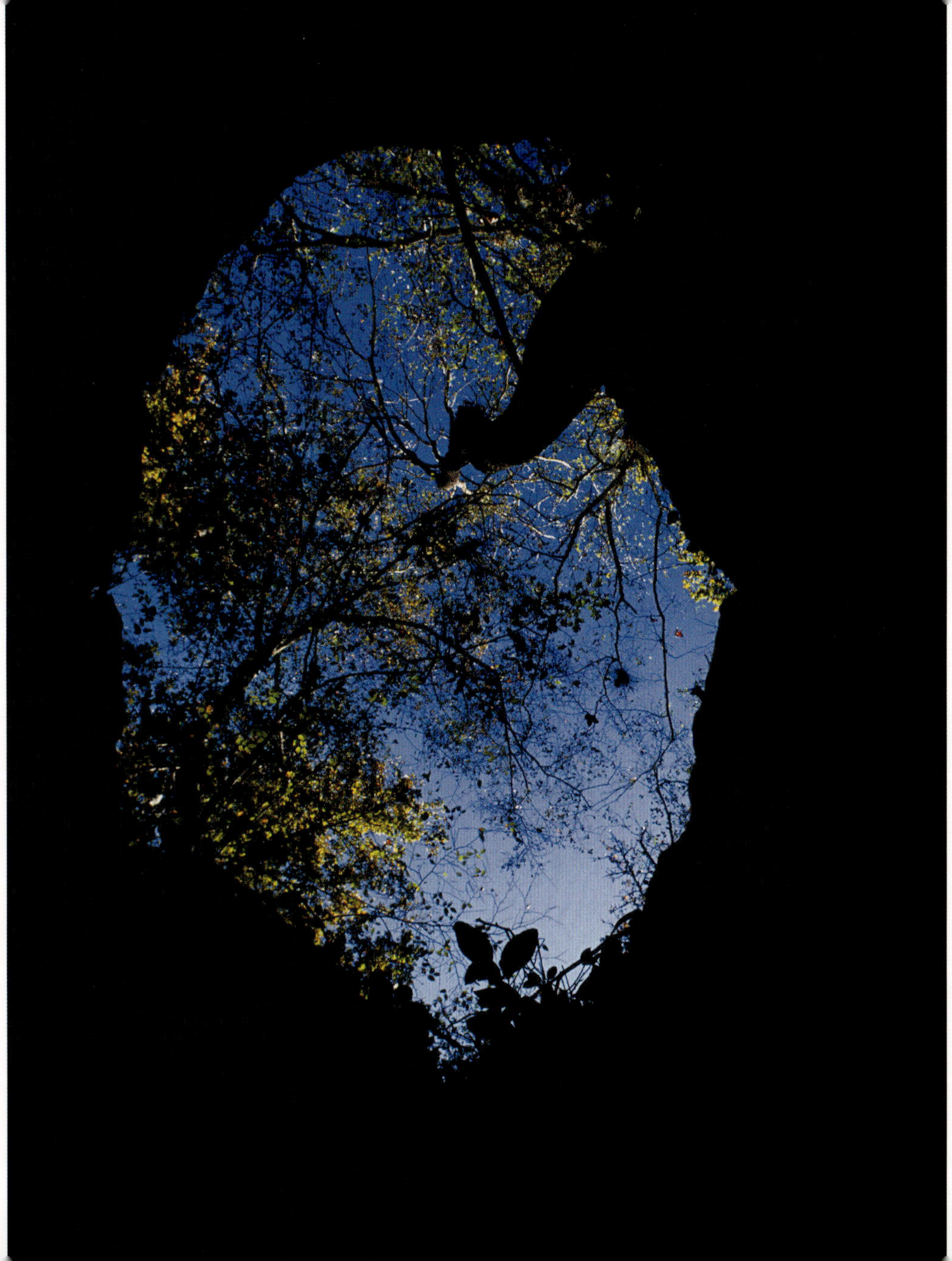

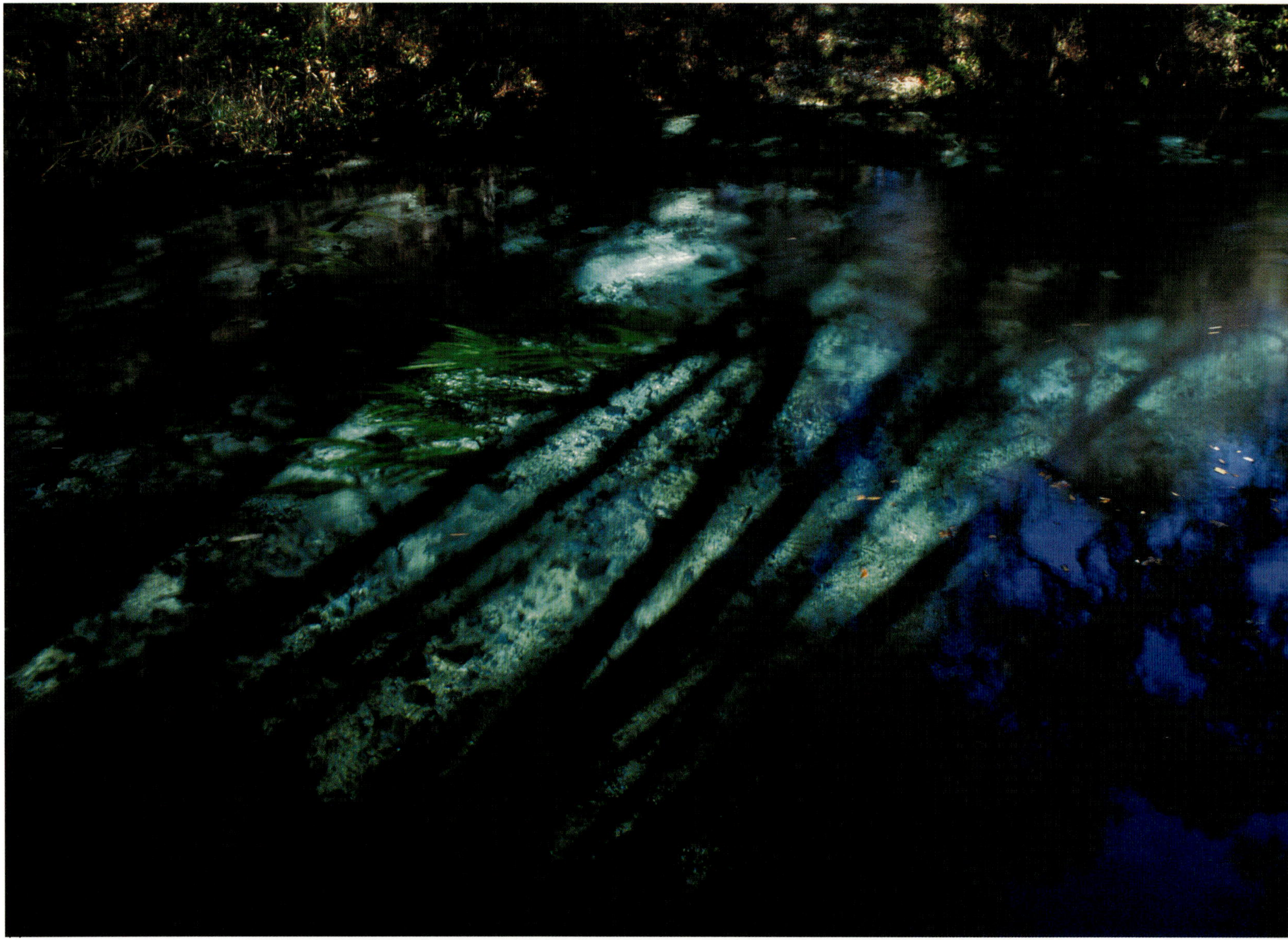

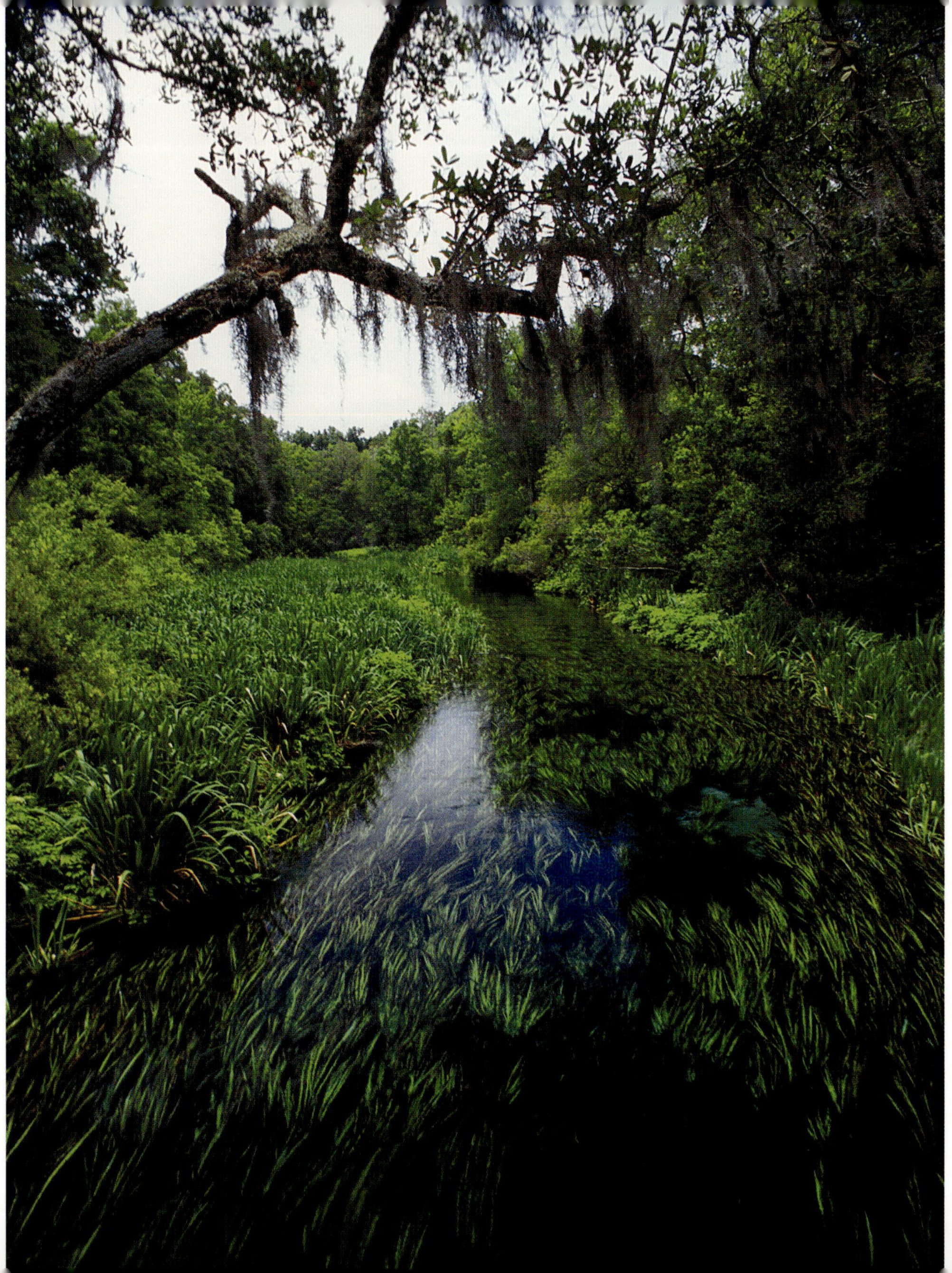

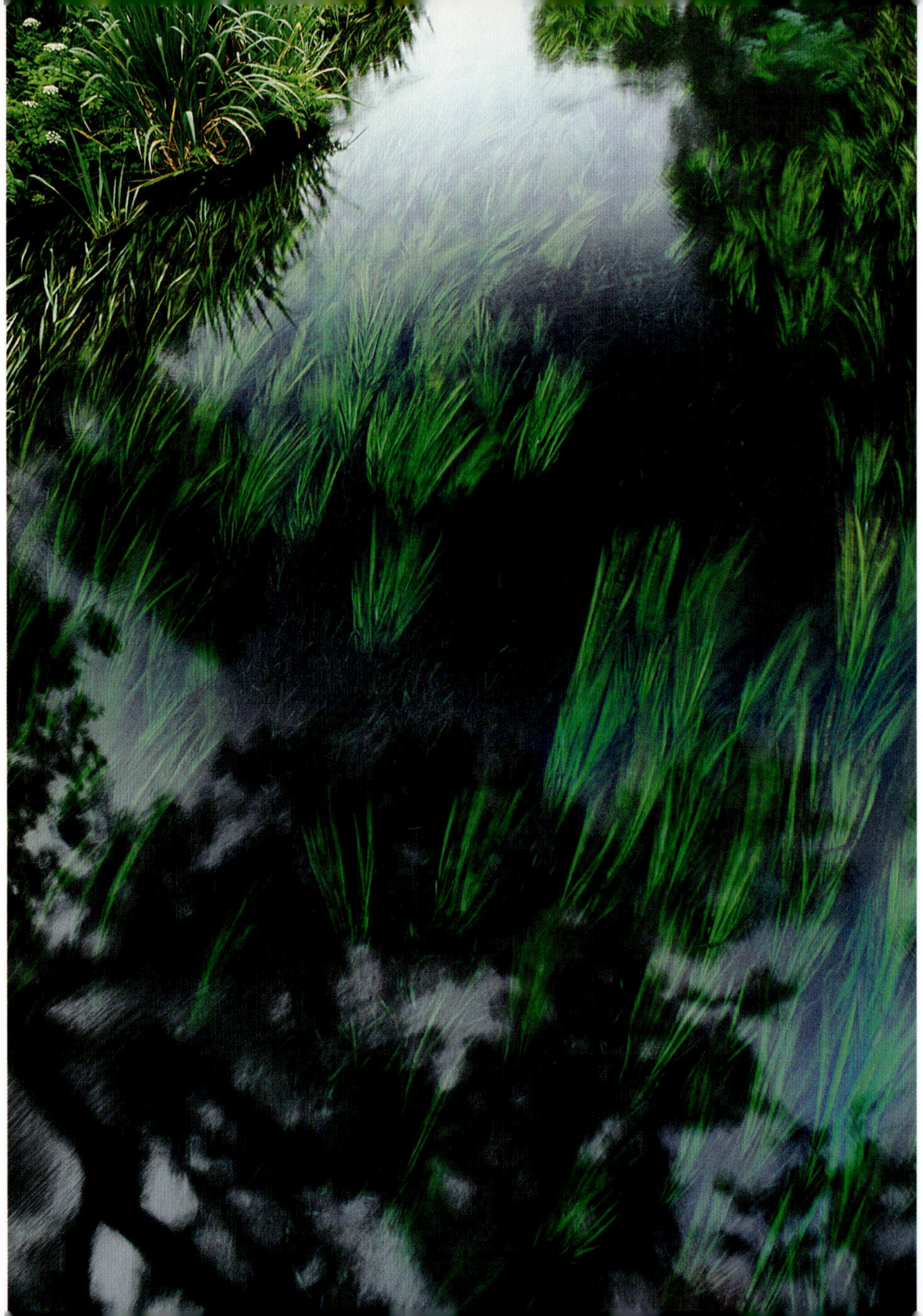

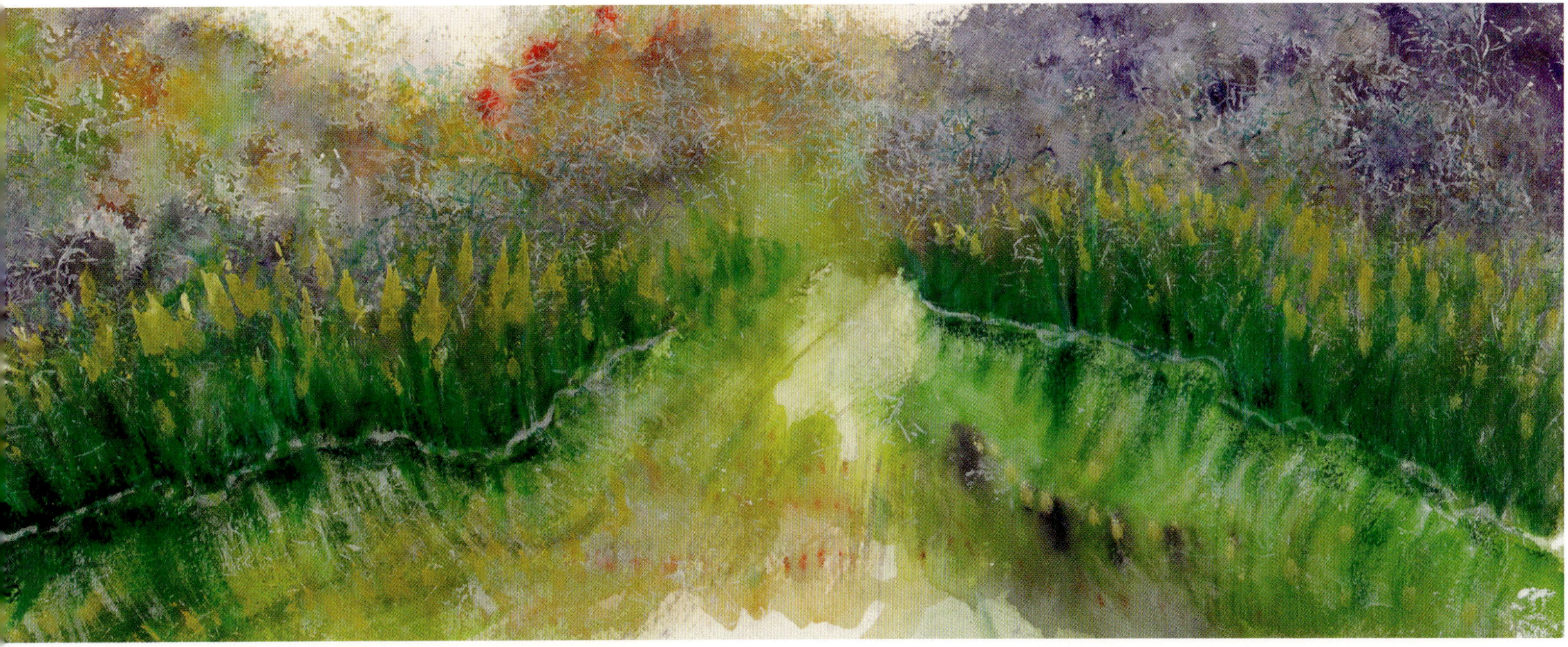

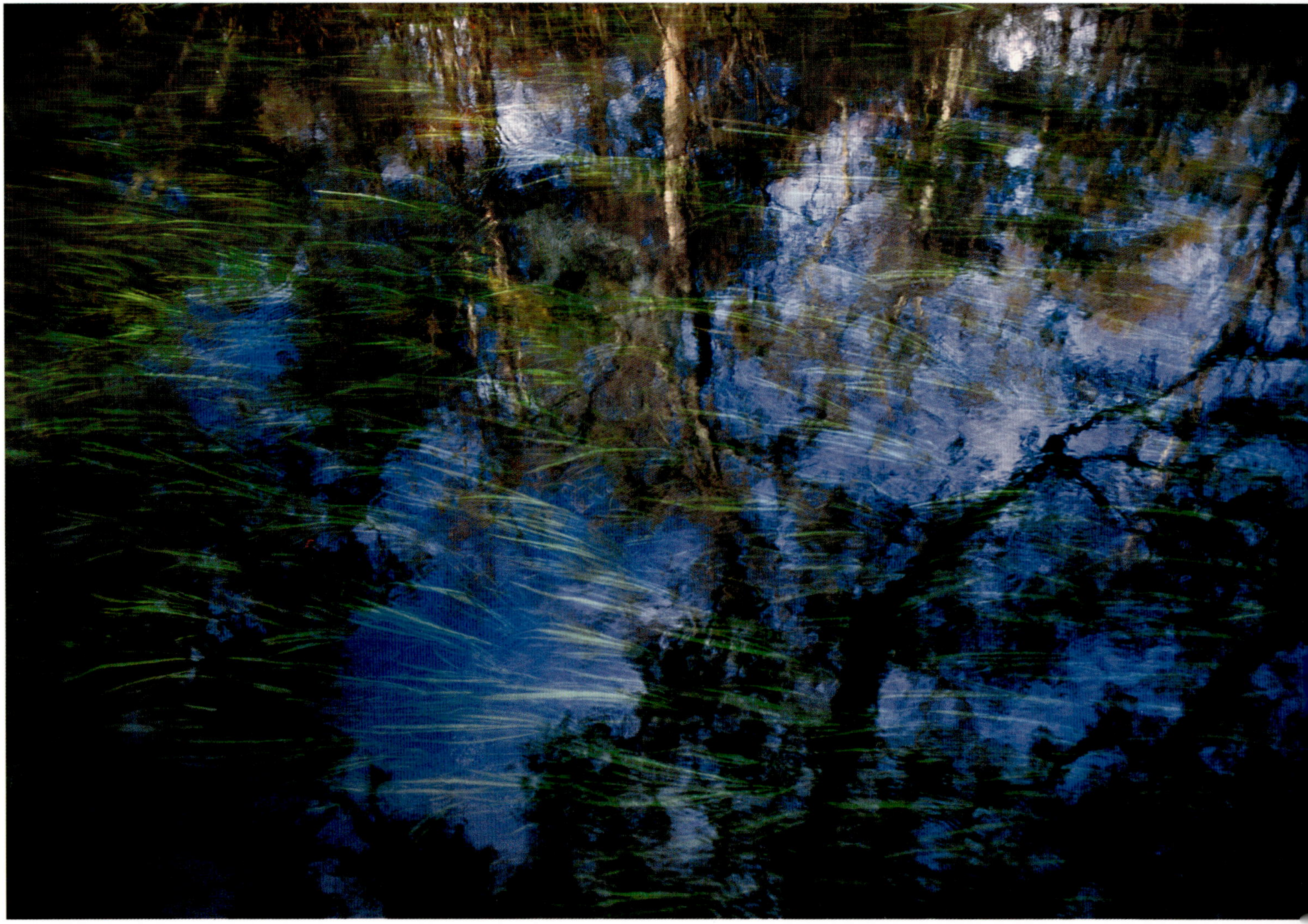

within the crystalline spring waters
flowing with the graceful movements
of vibrantly green grasses
the memory of Ichetucknee's untouched purity pervades

the restorative powers of cleansing and healing
still exist for this resilient river

the underlying beauty of her sacred indomitable spirit
touches all who know and come to understand her

the magical essence of this imperiled river
alive in her undying spirit
the ultimate hope of rejuvenescence

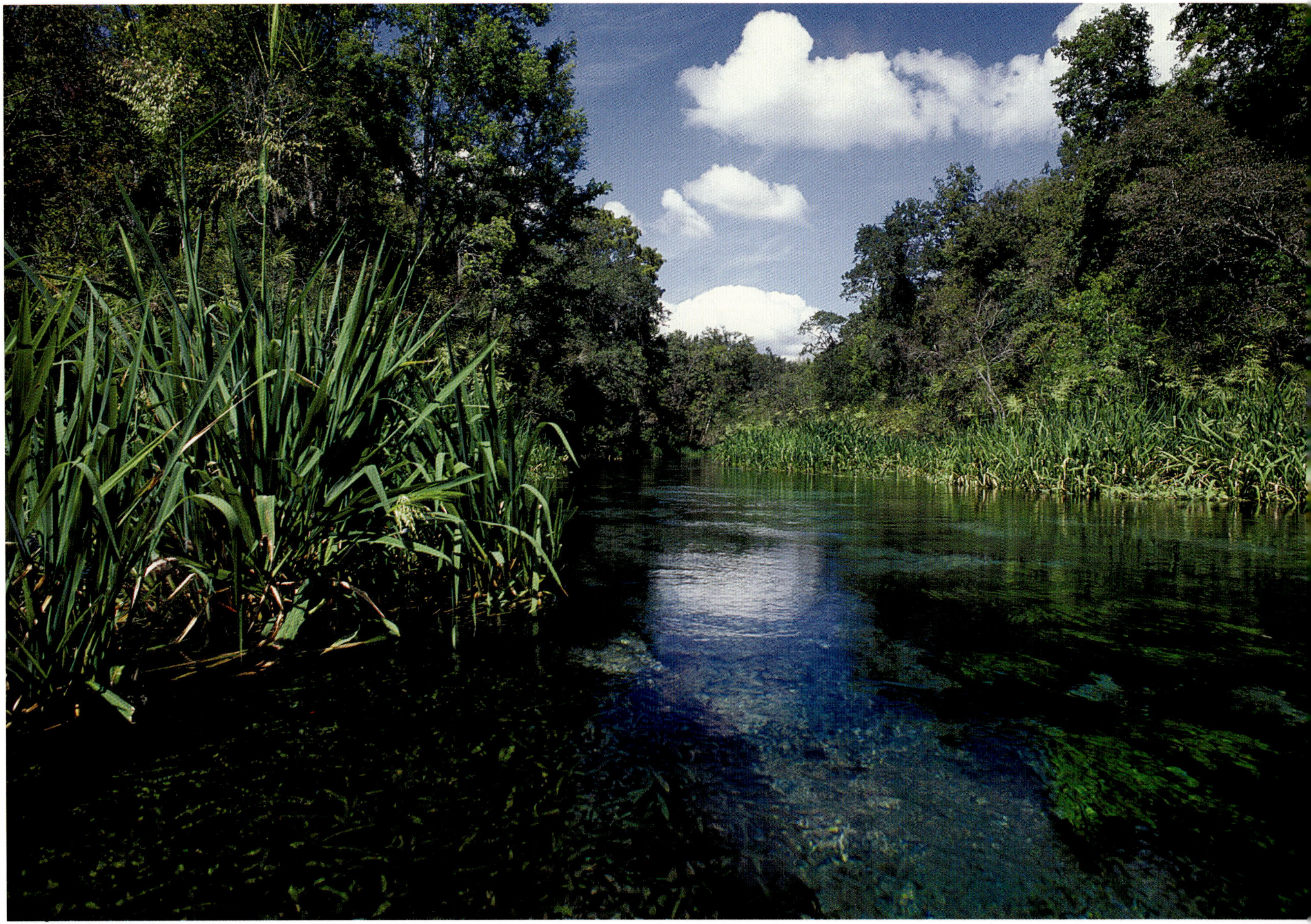

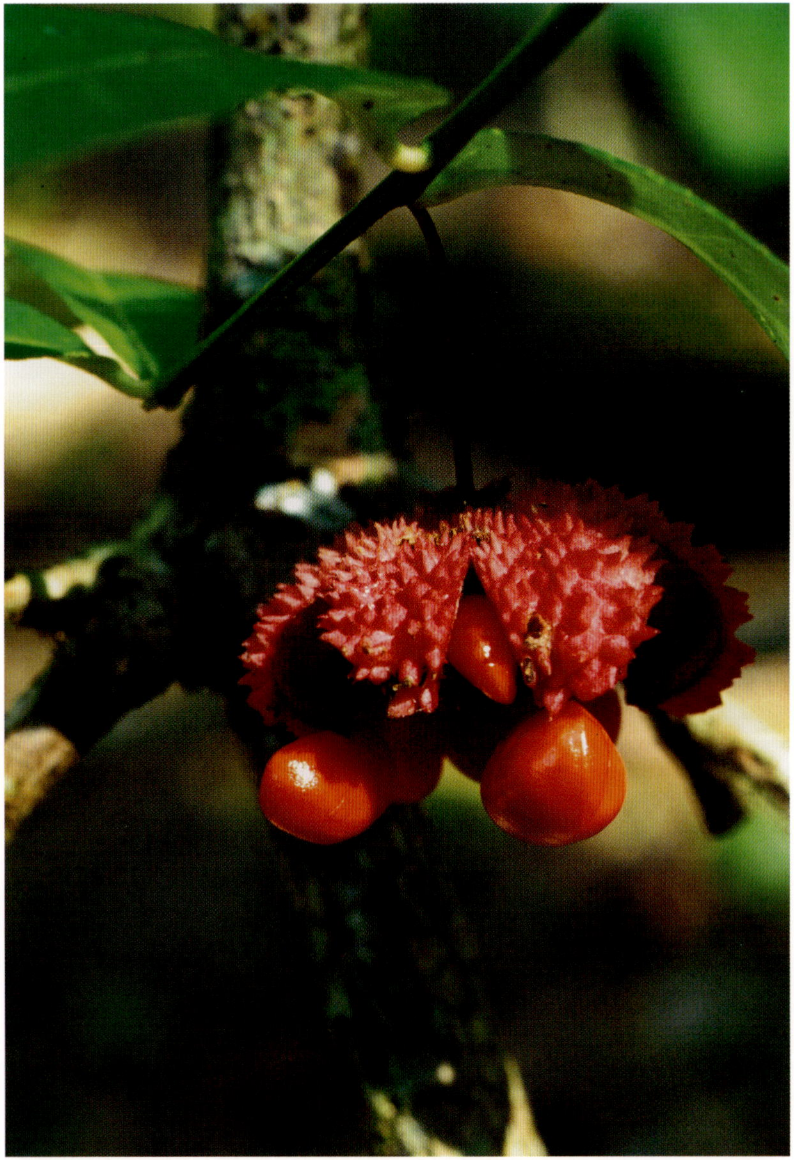

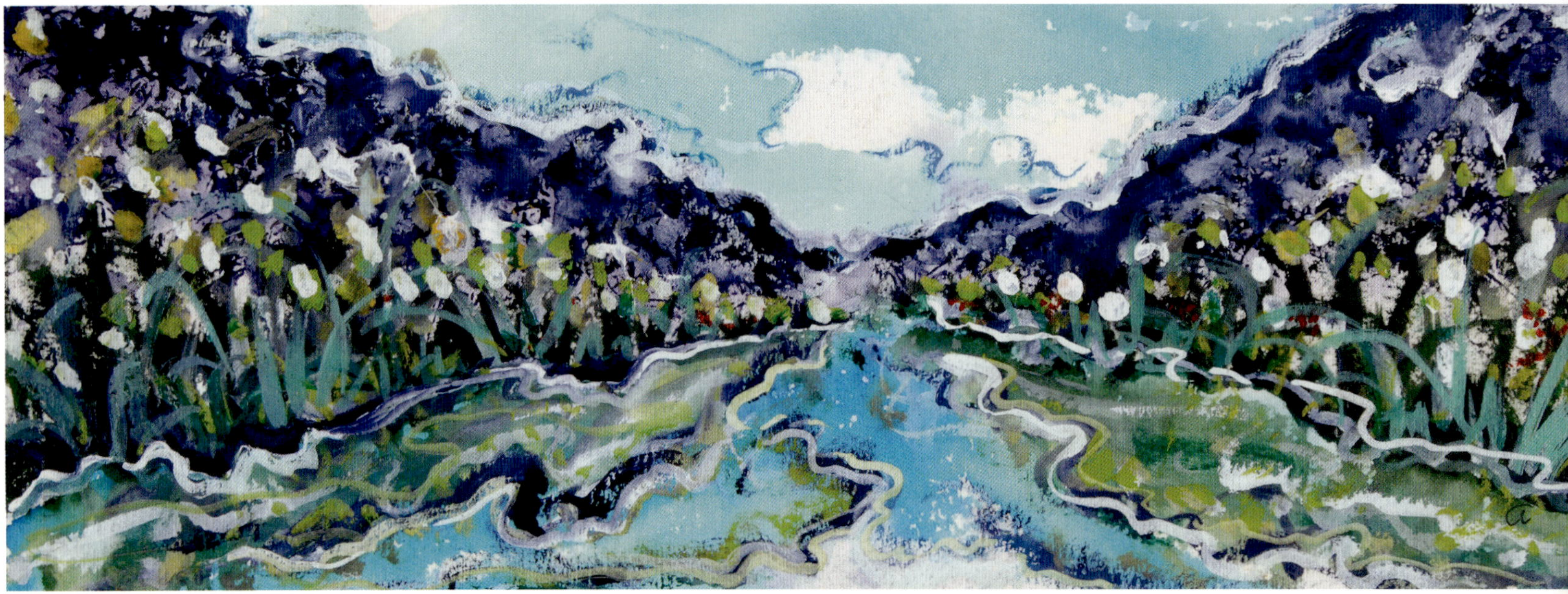

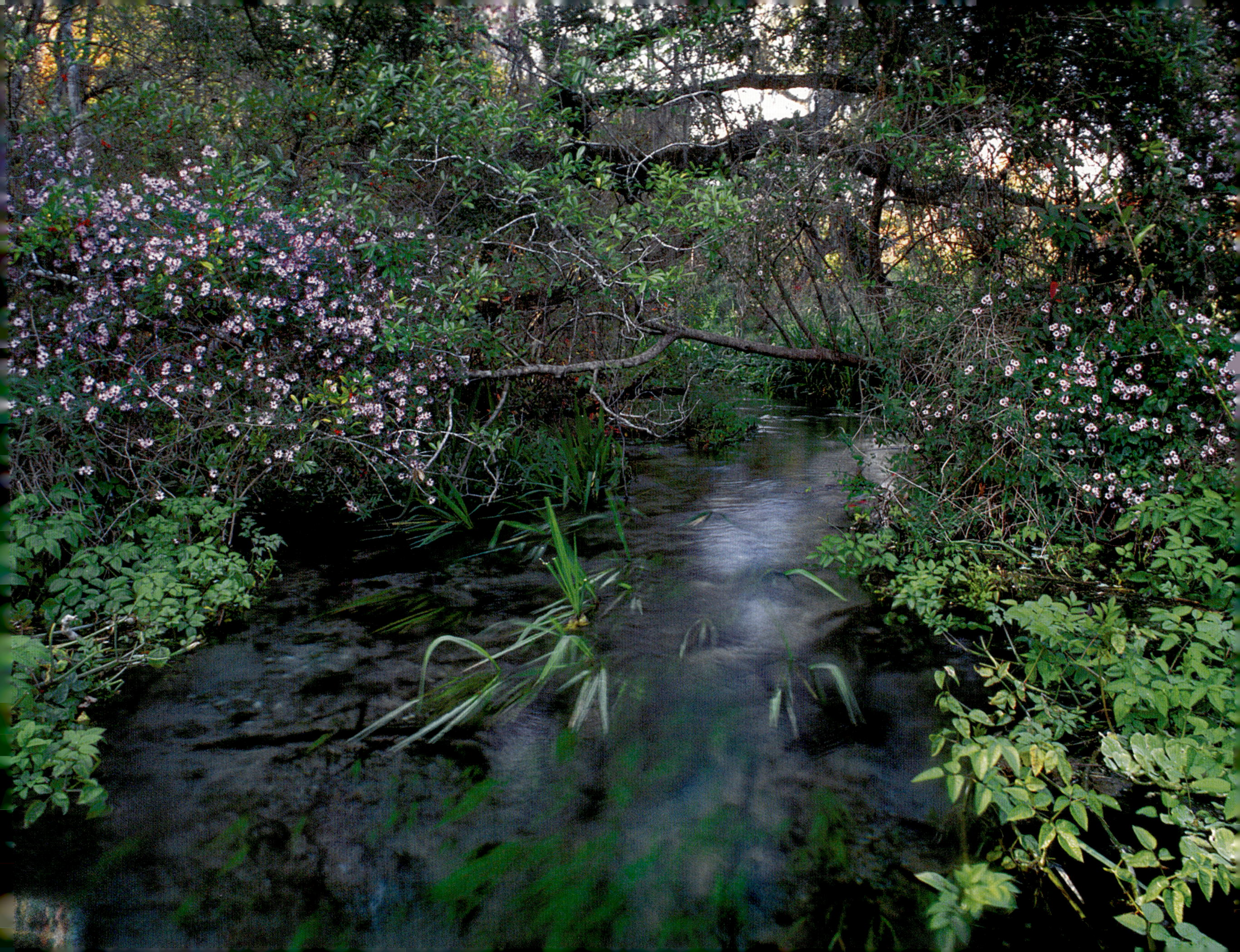

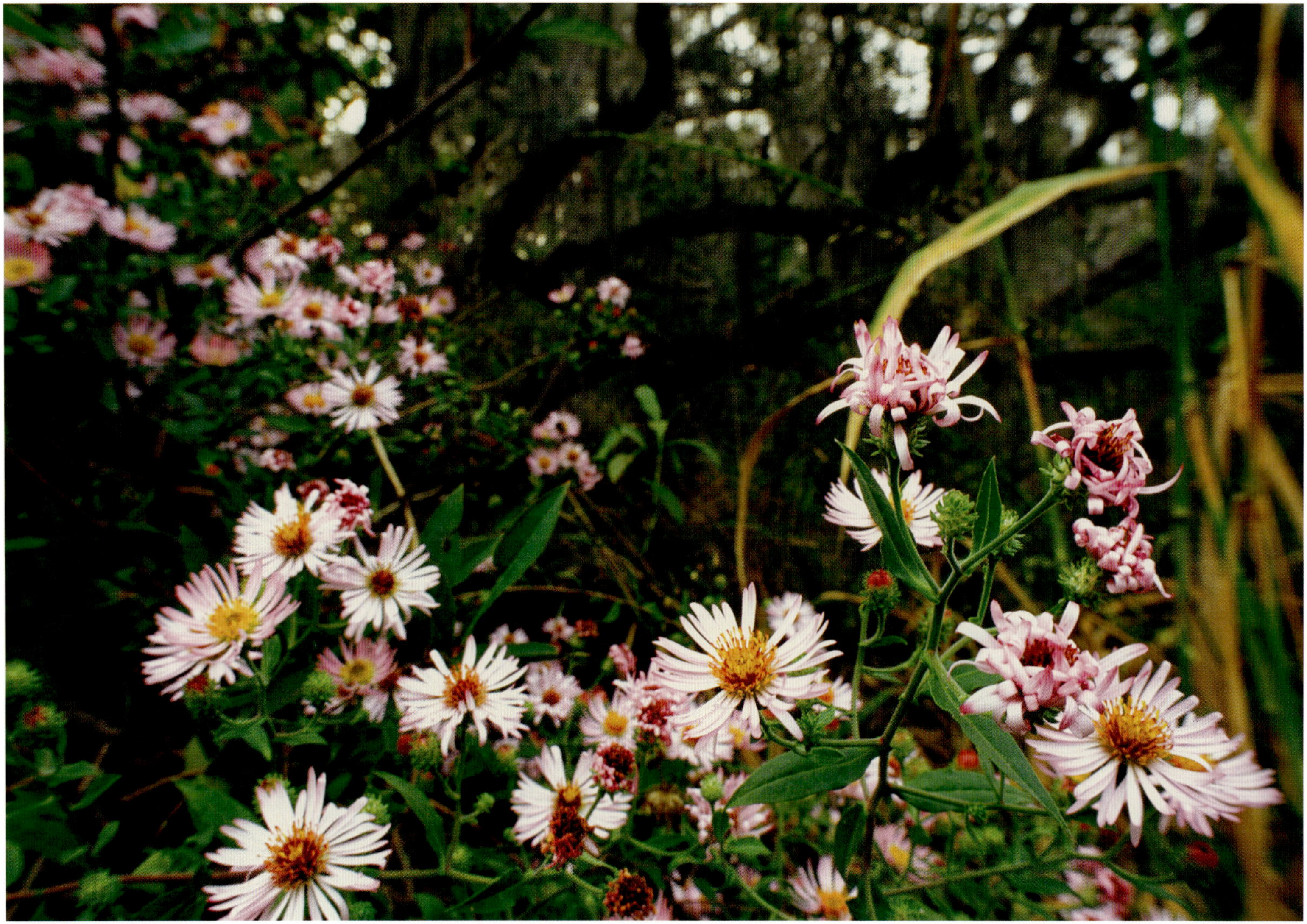

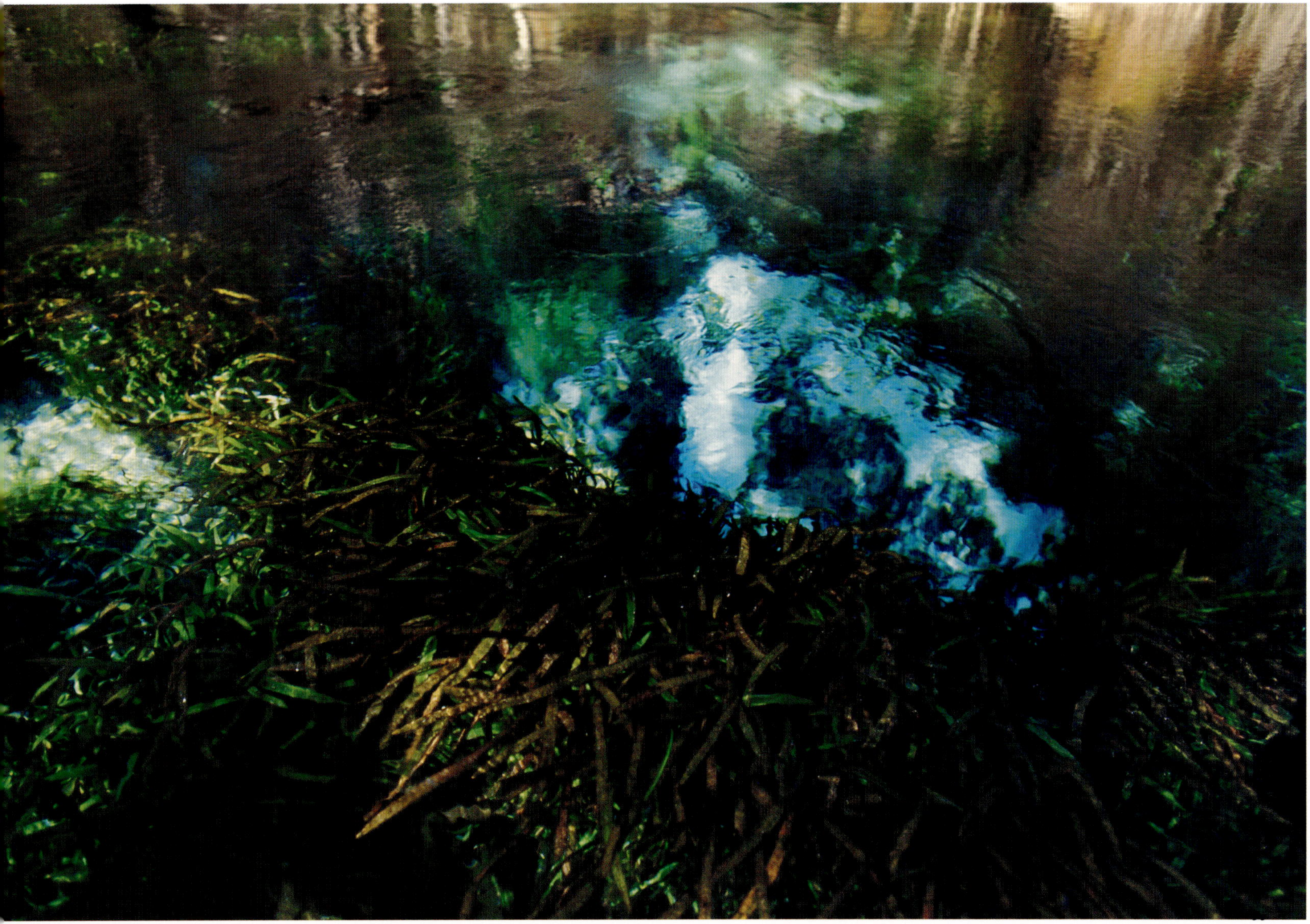

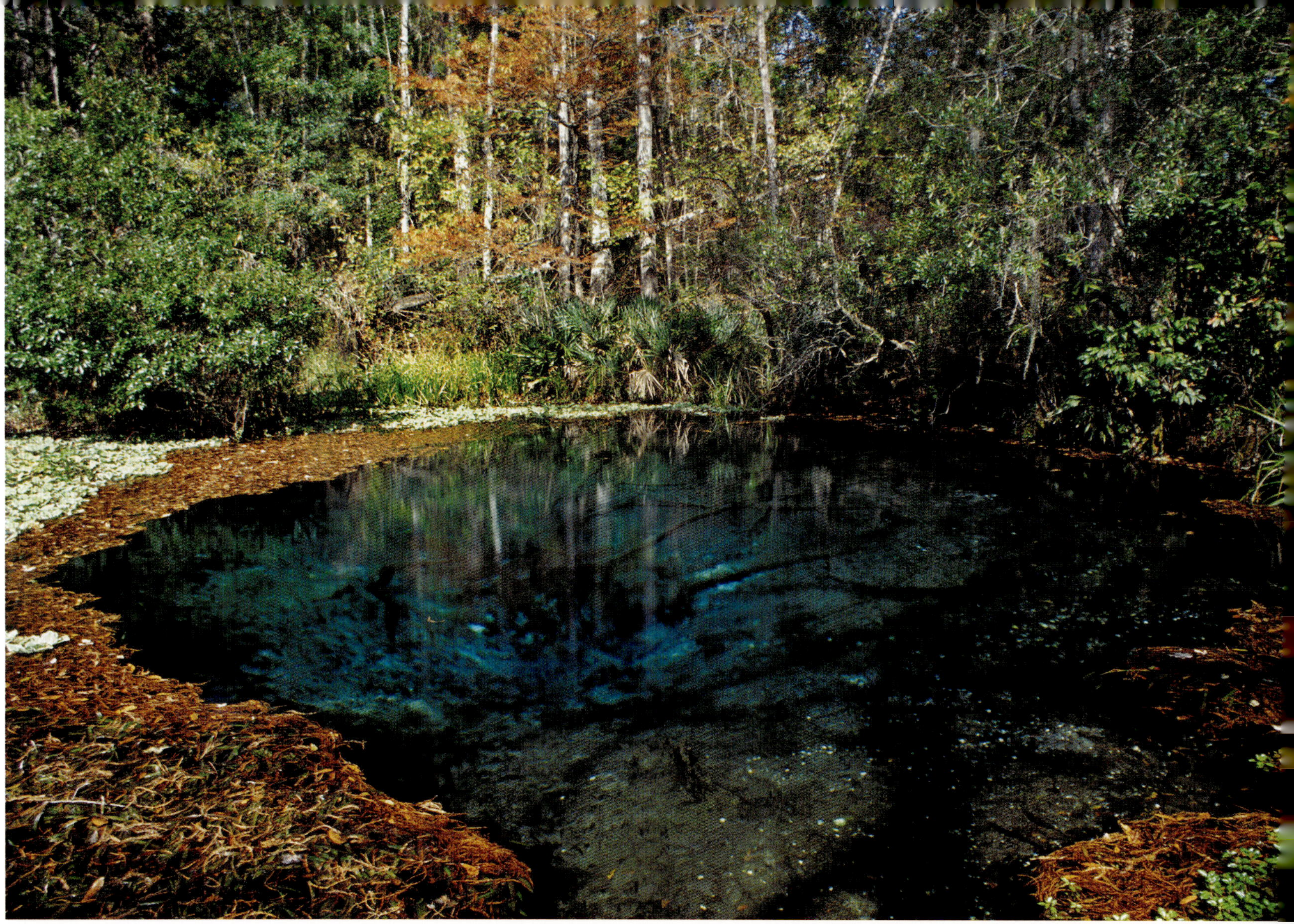

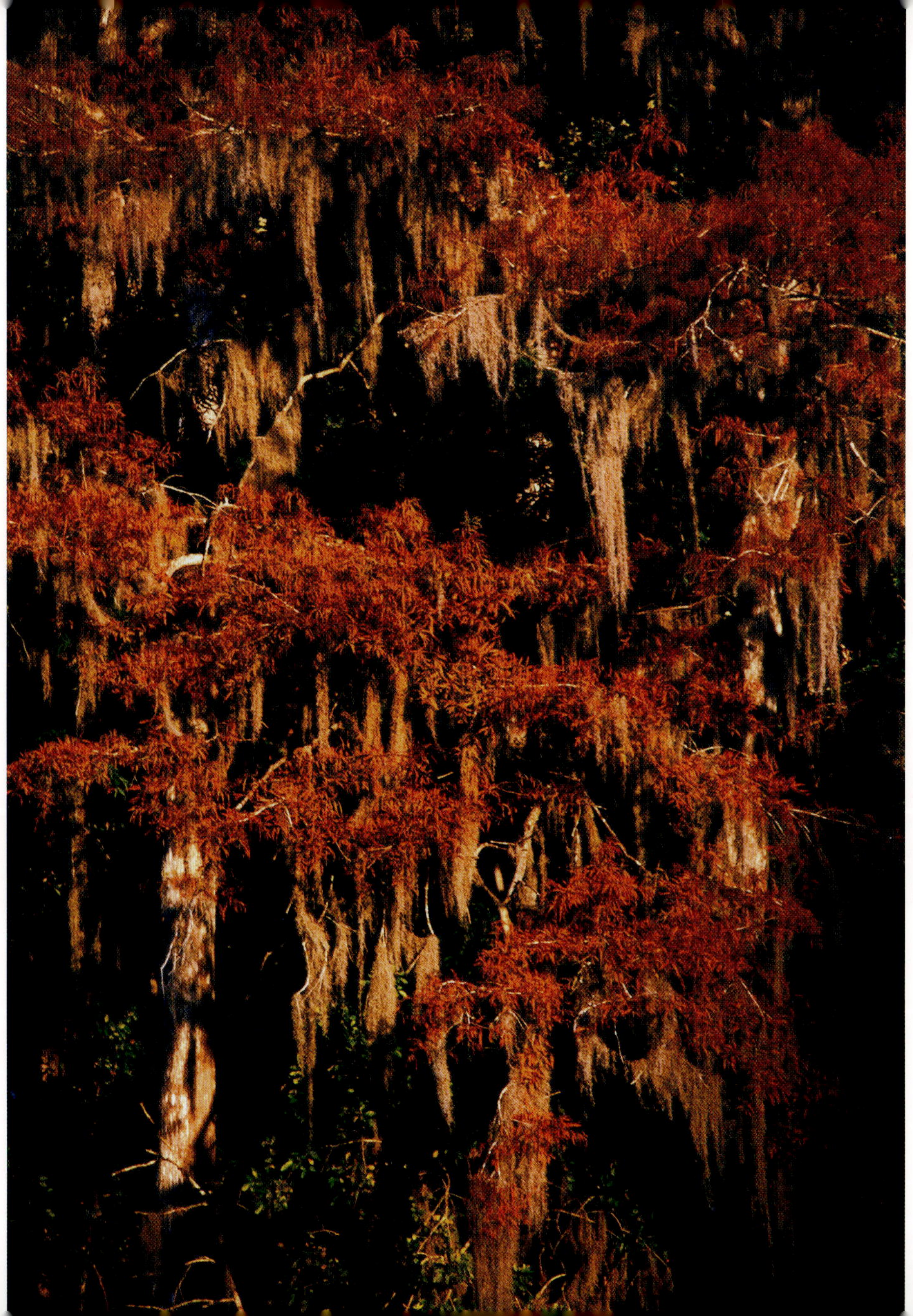

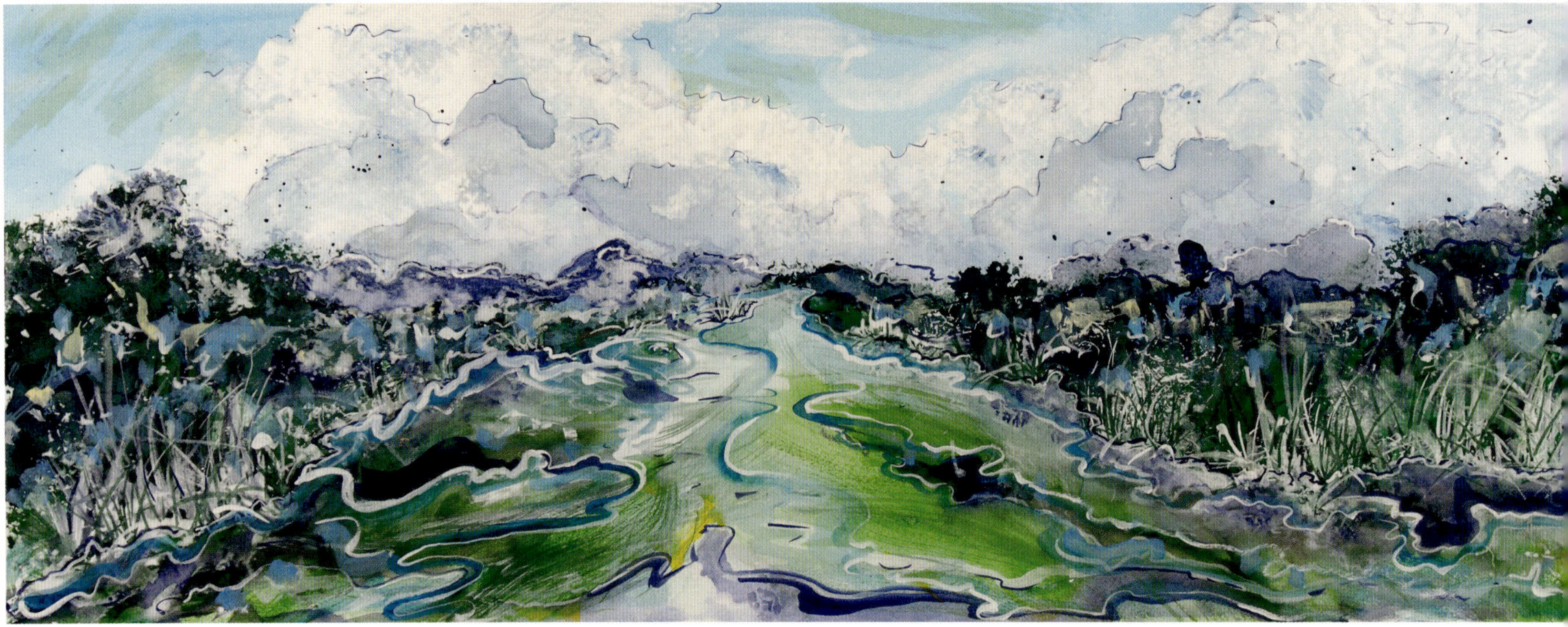

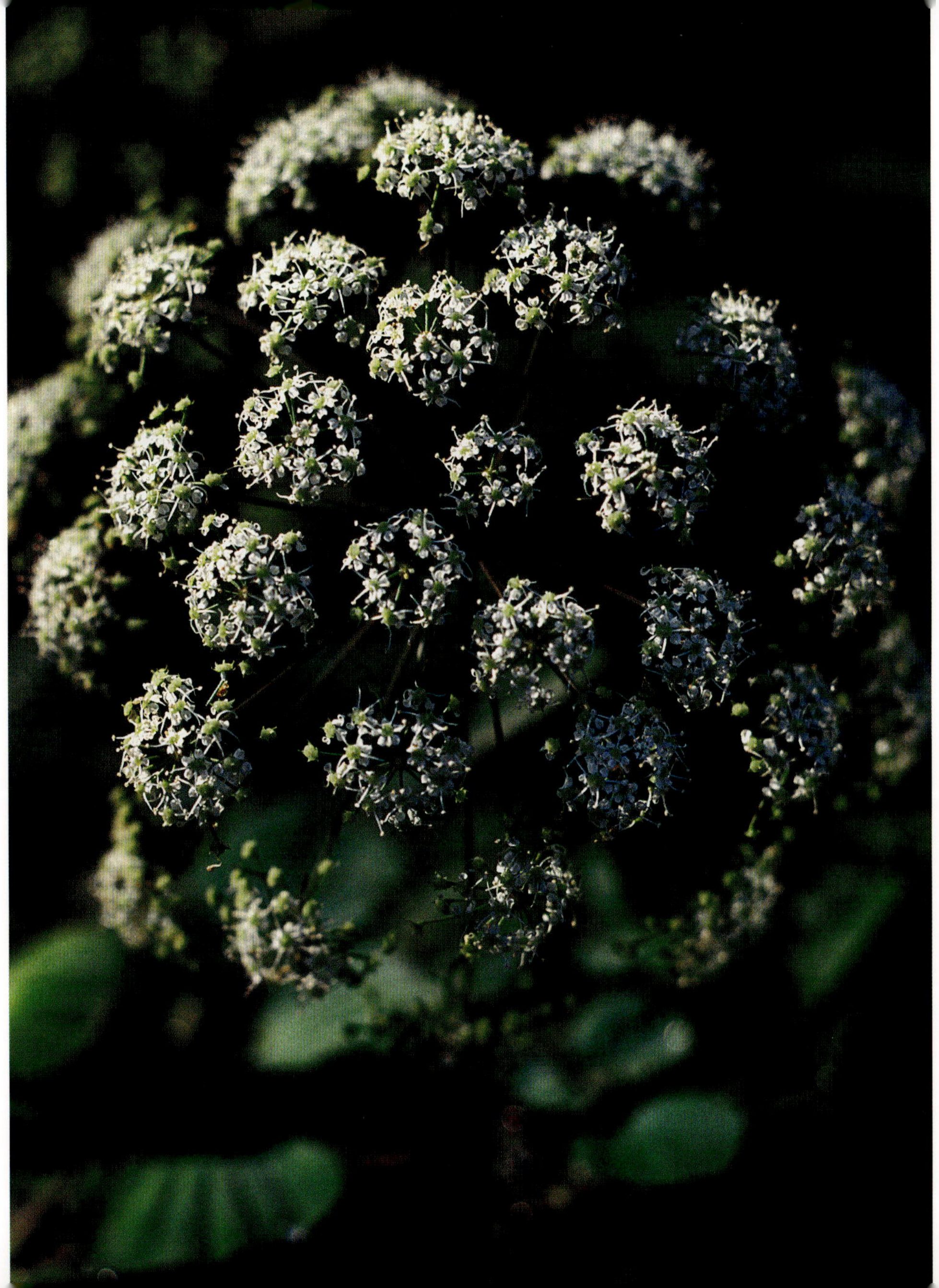

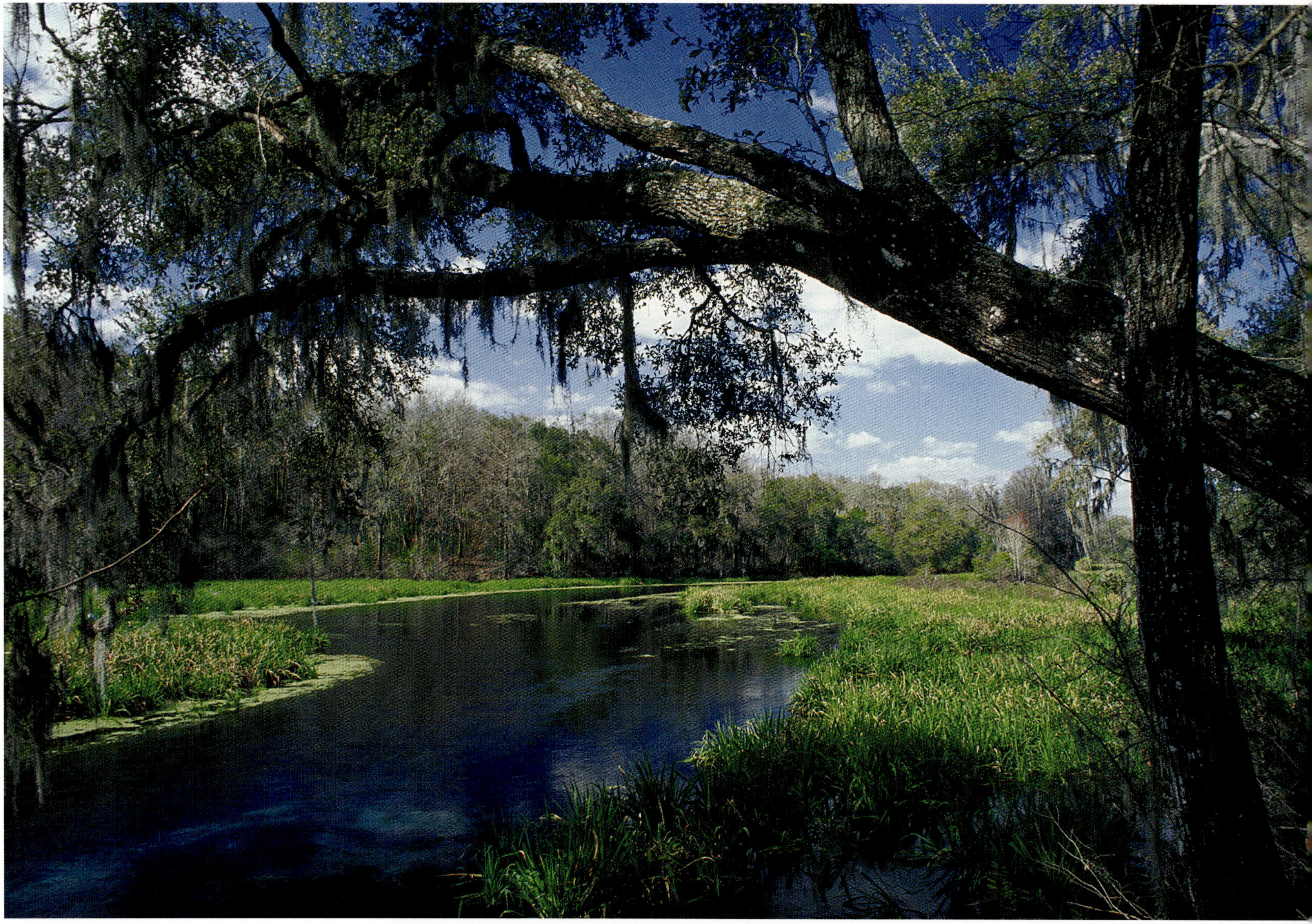

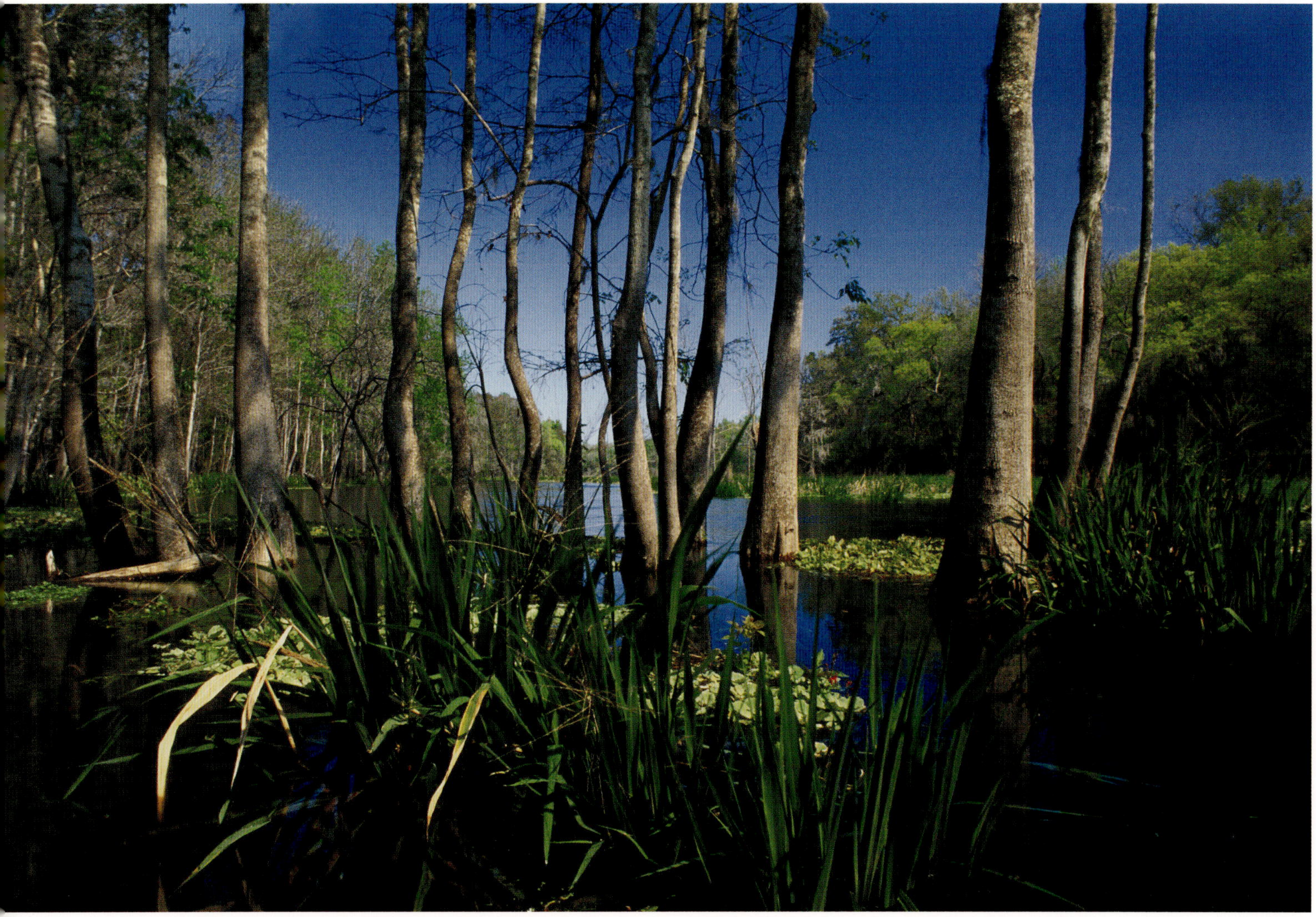

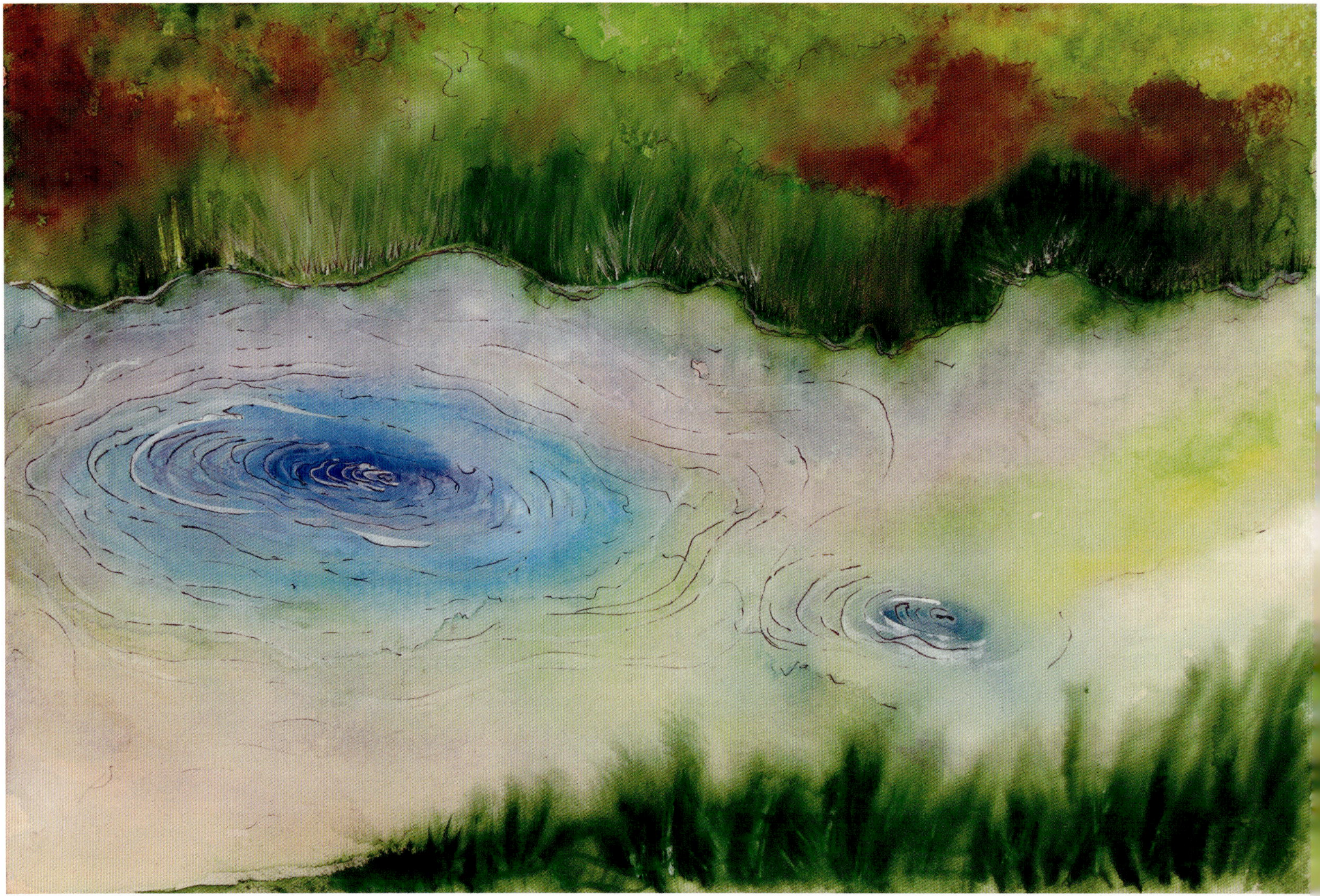

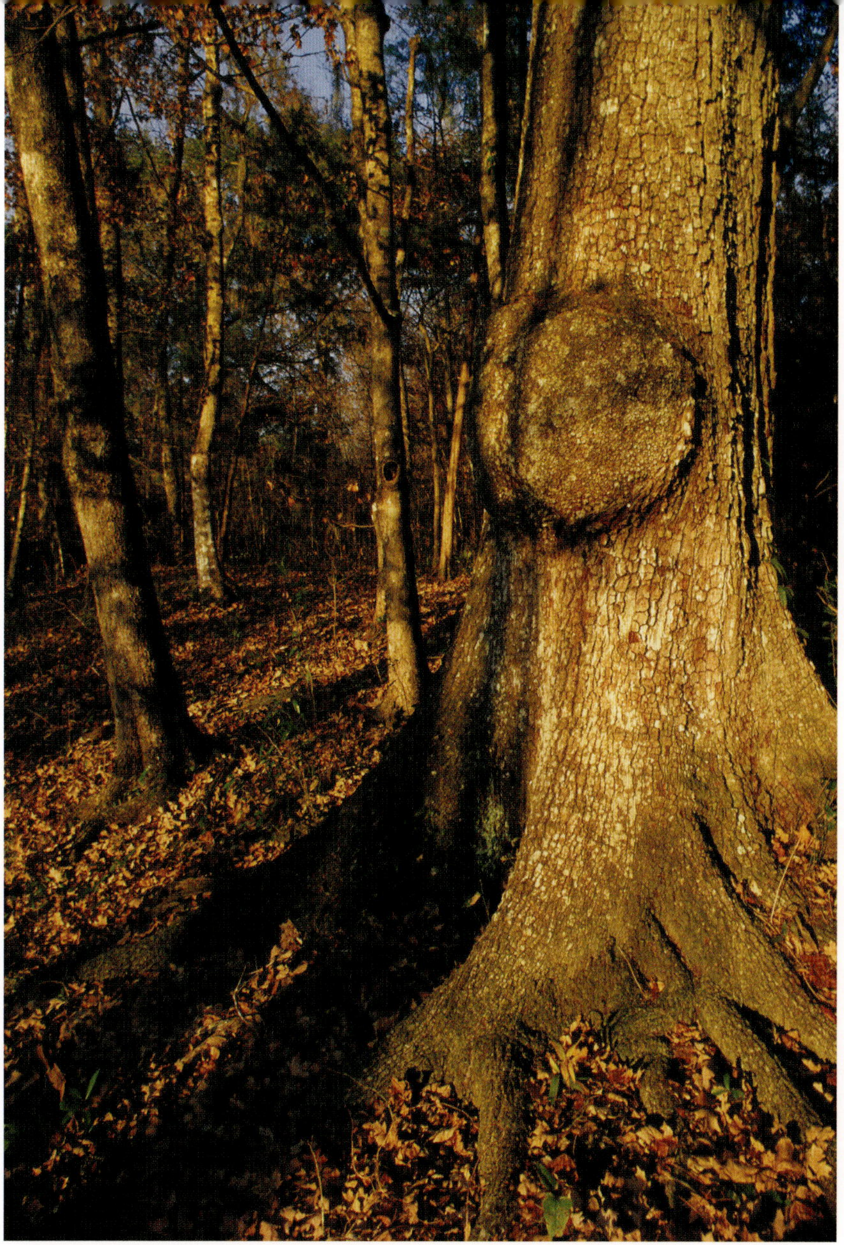

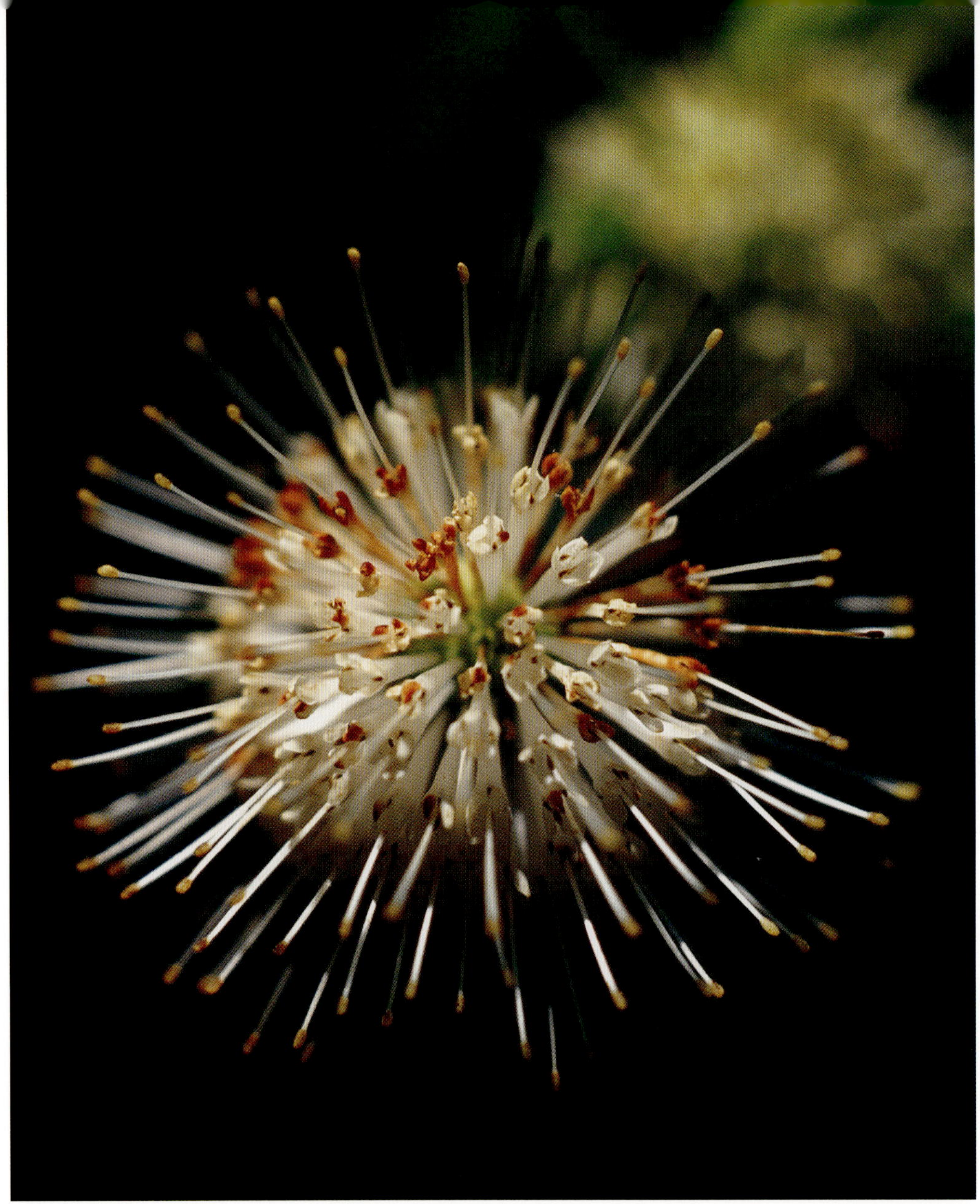

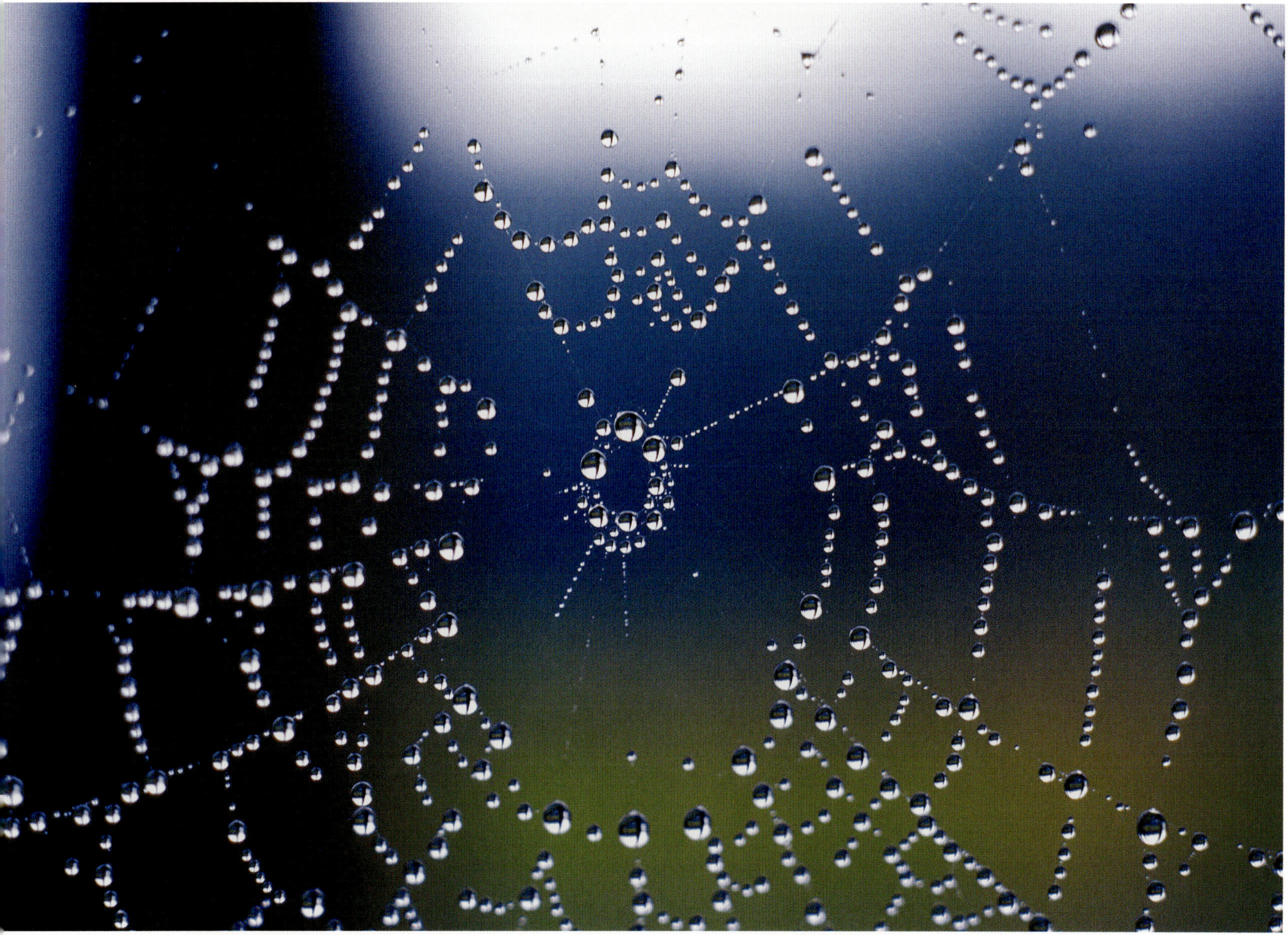

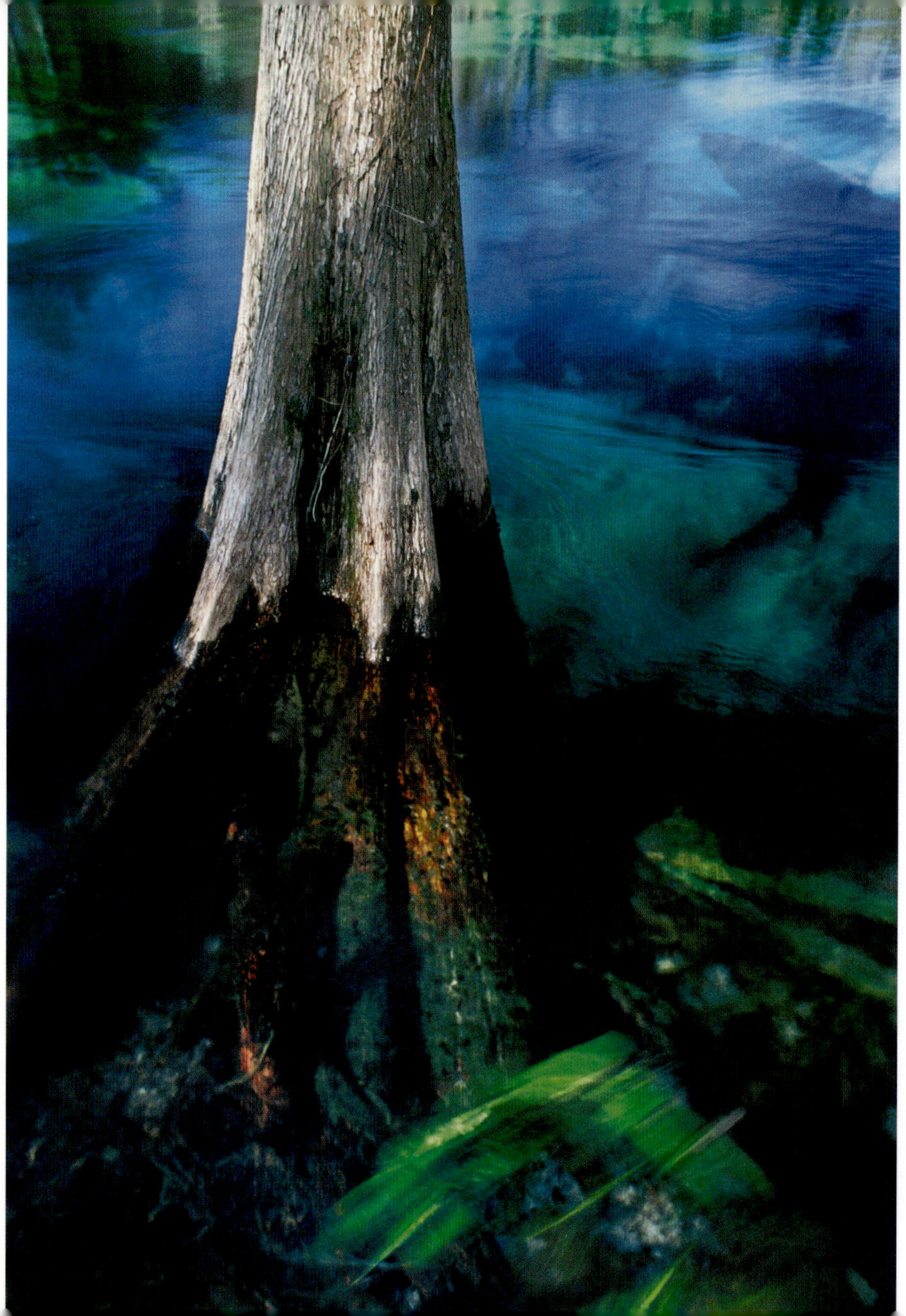

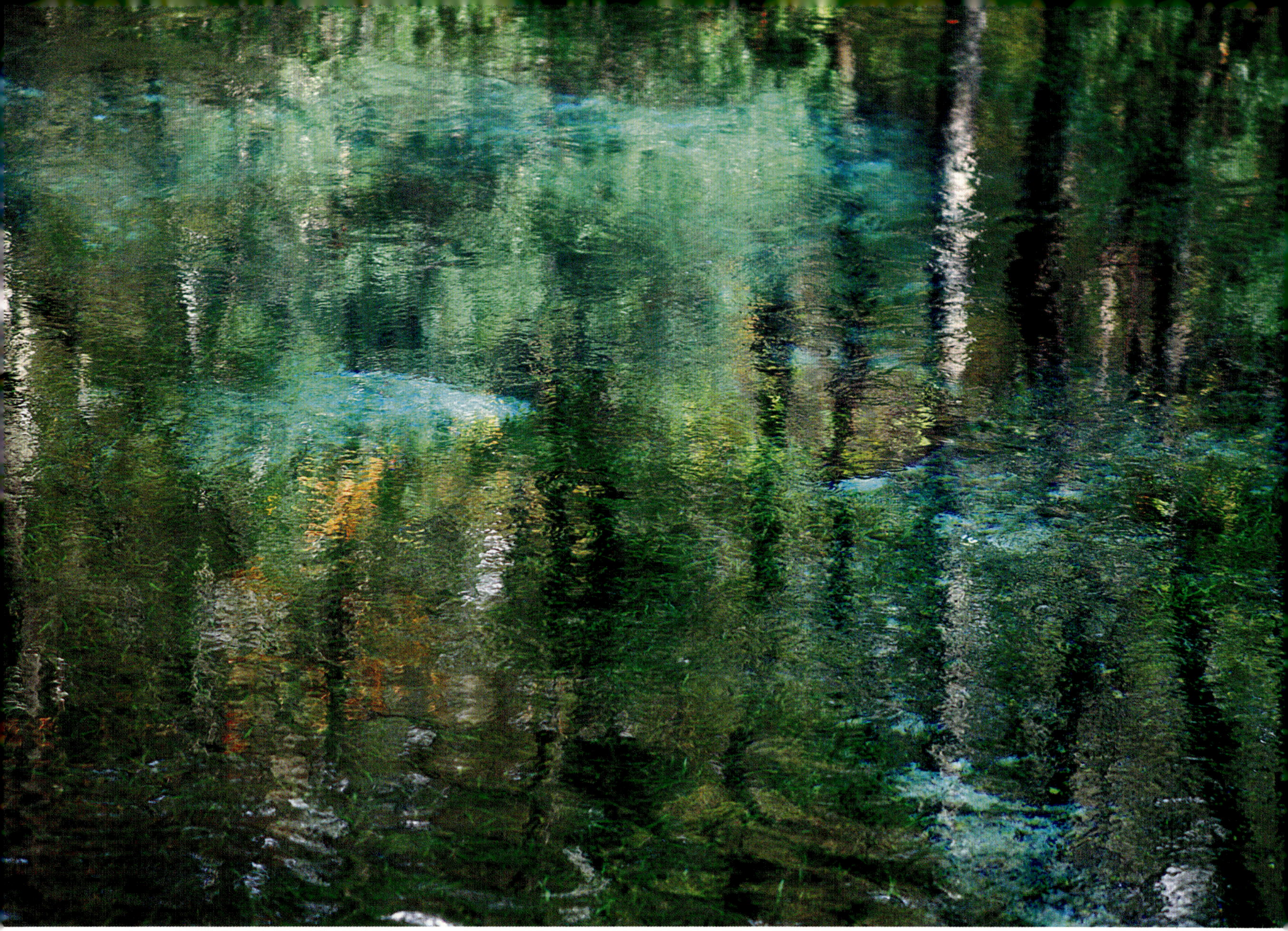

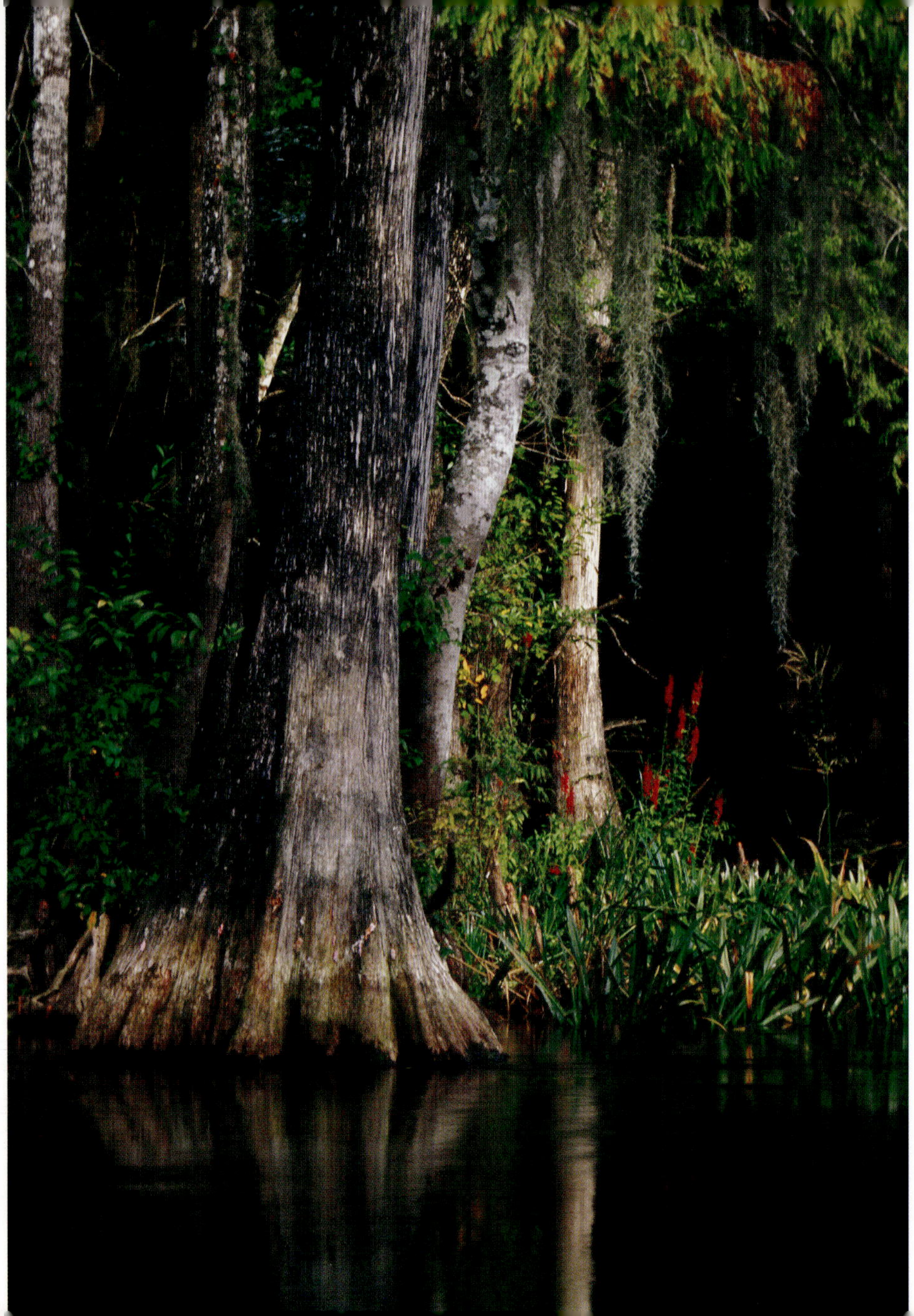

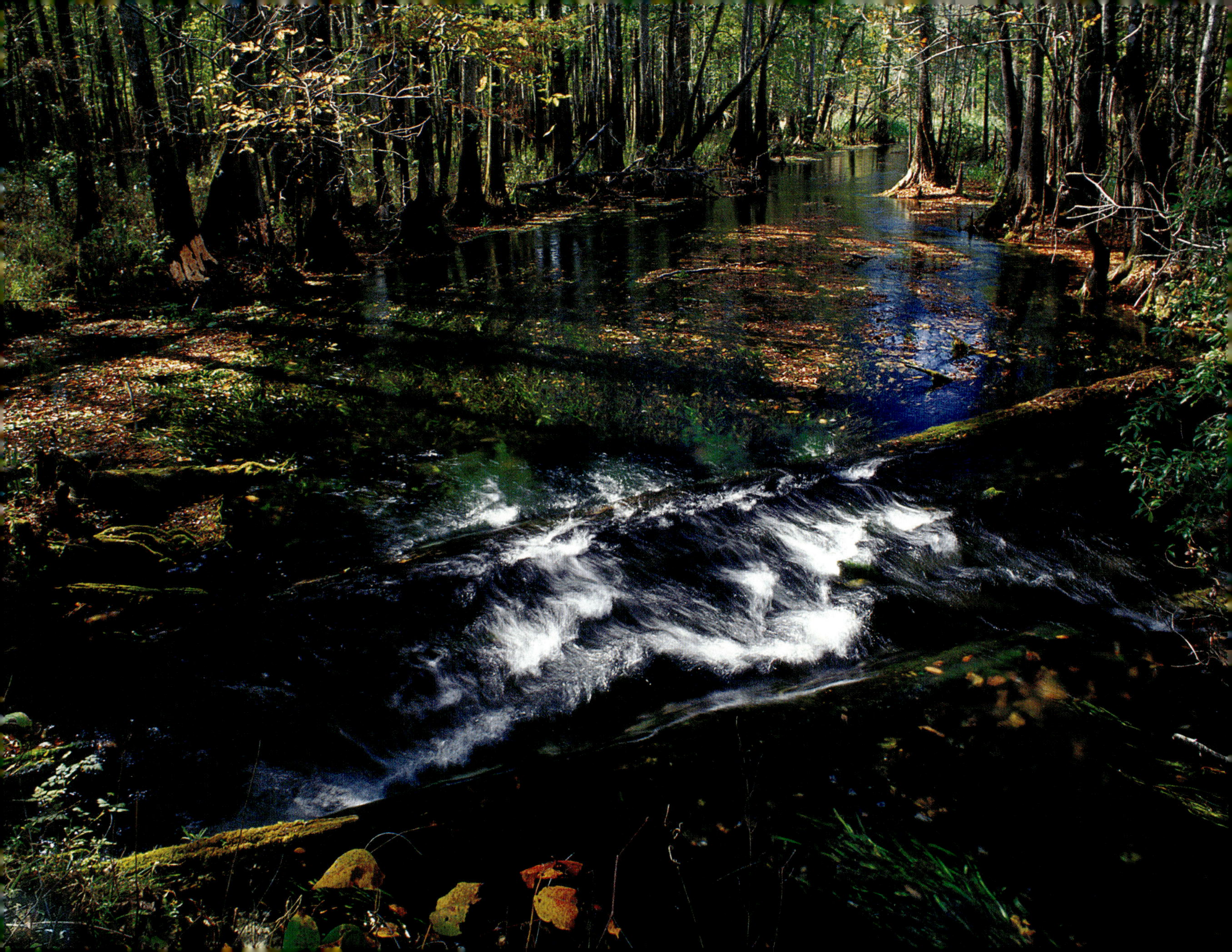

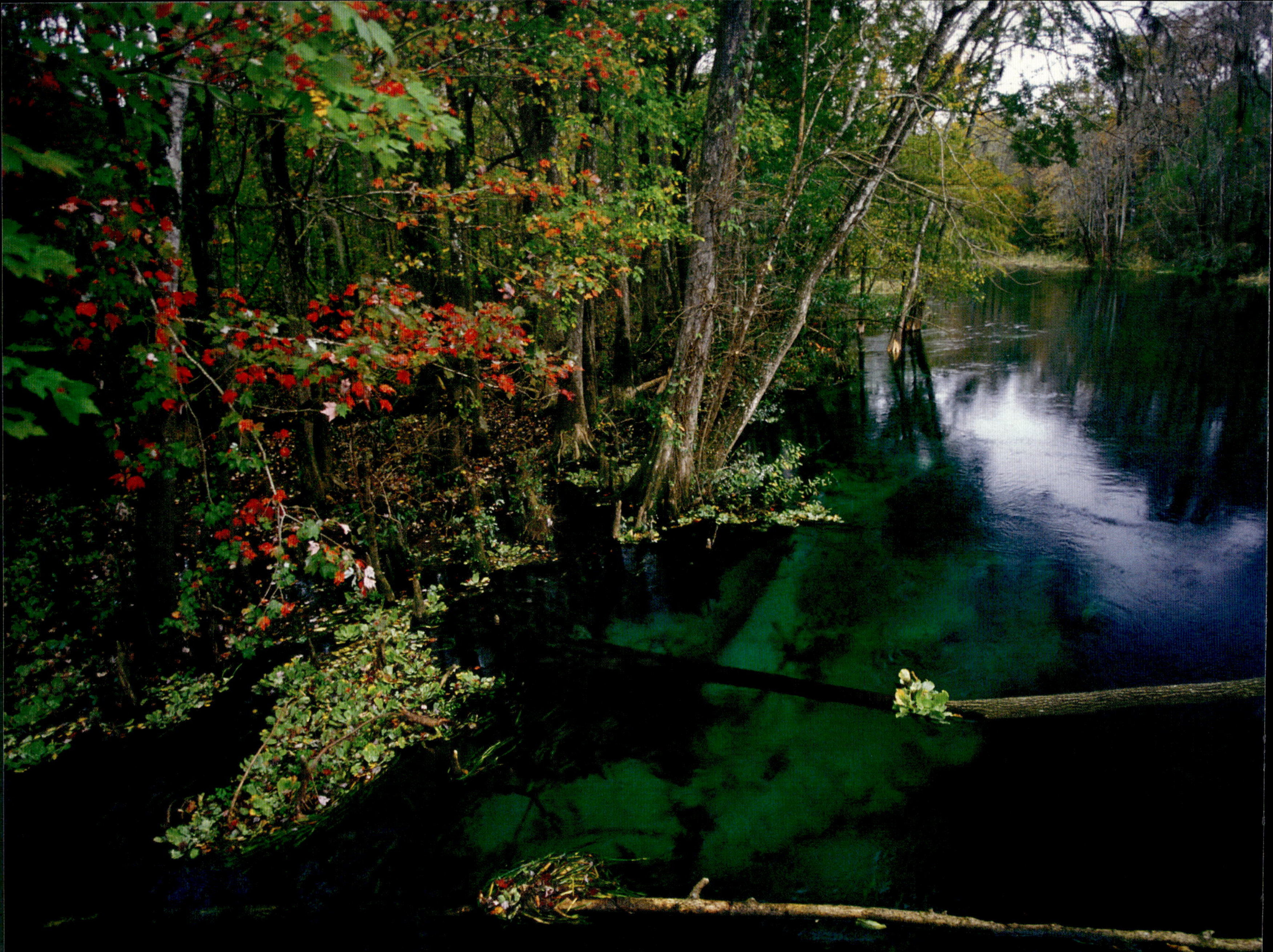

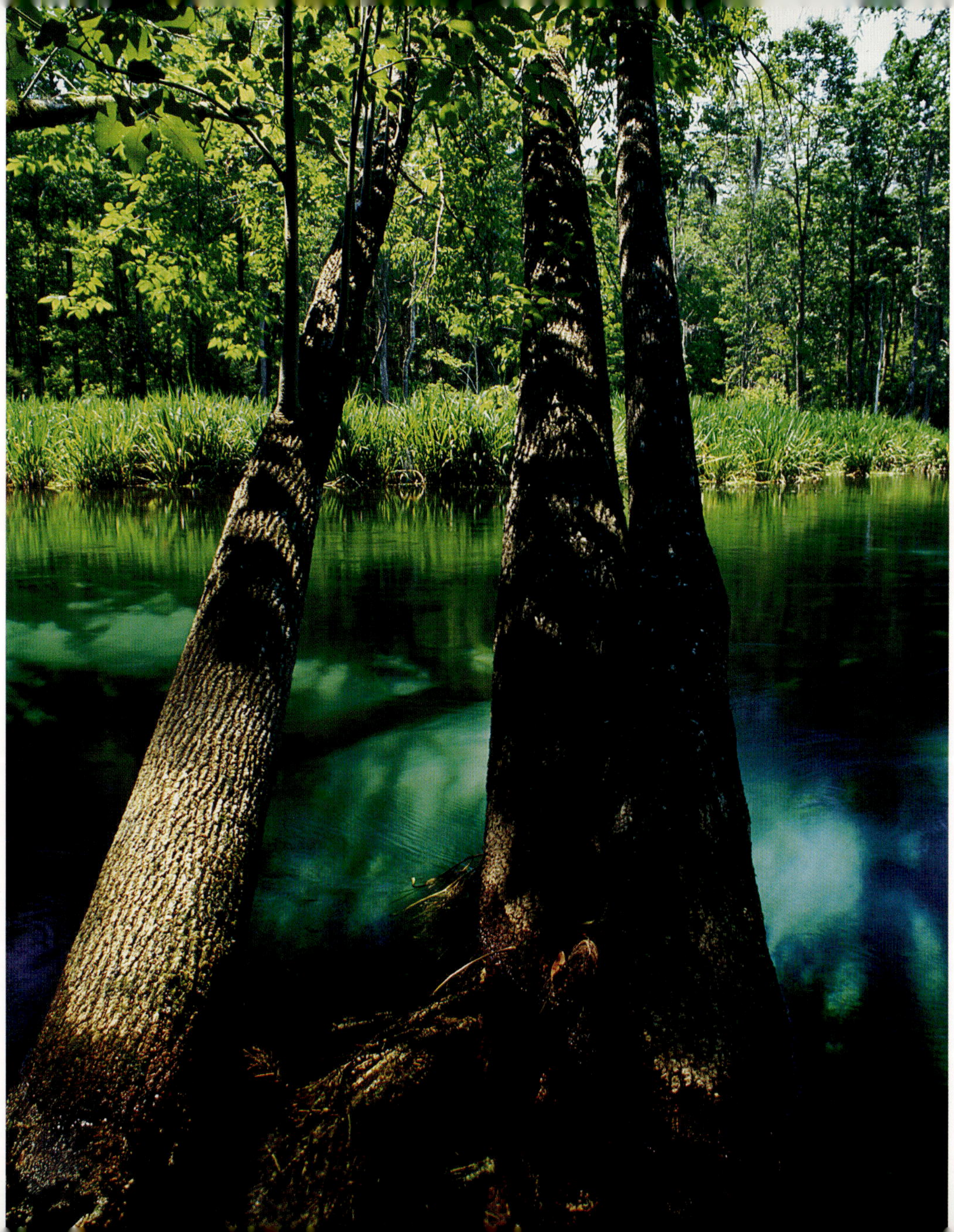

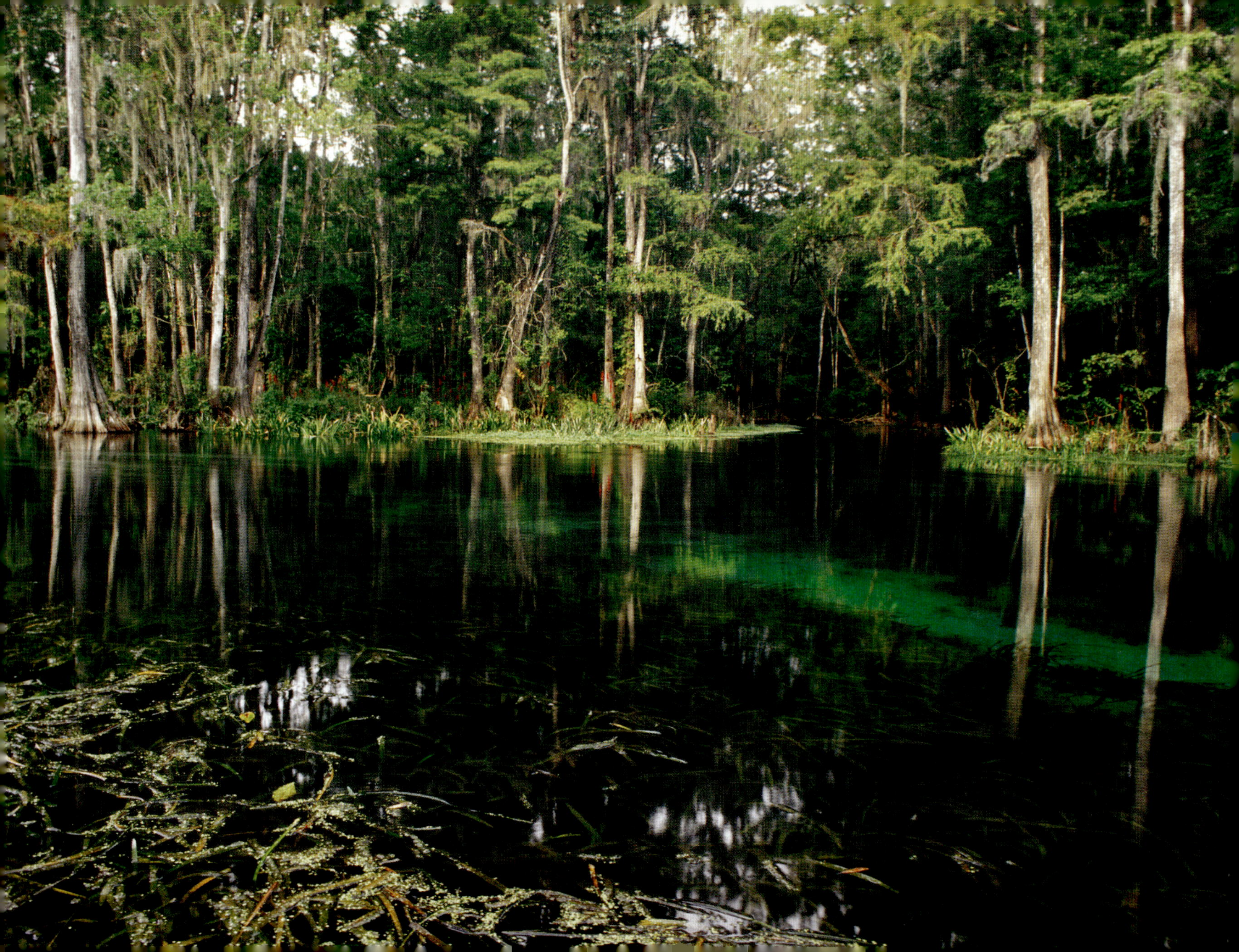

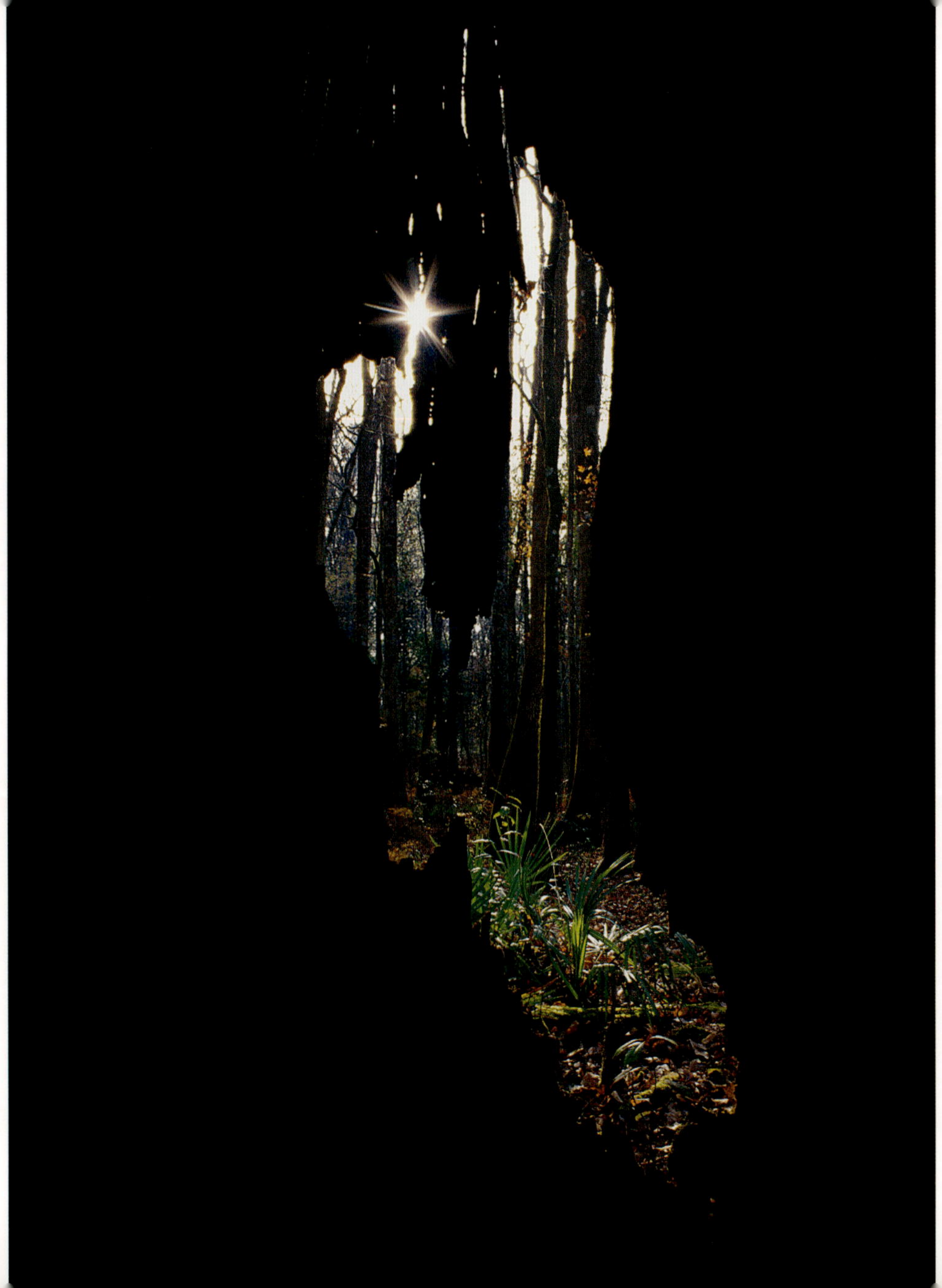

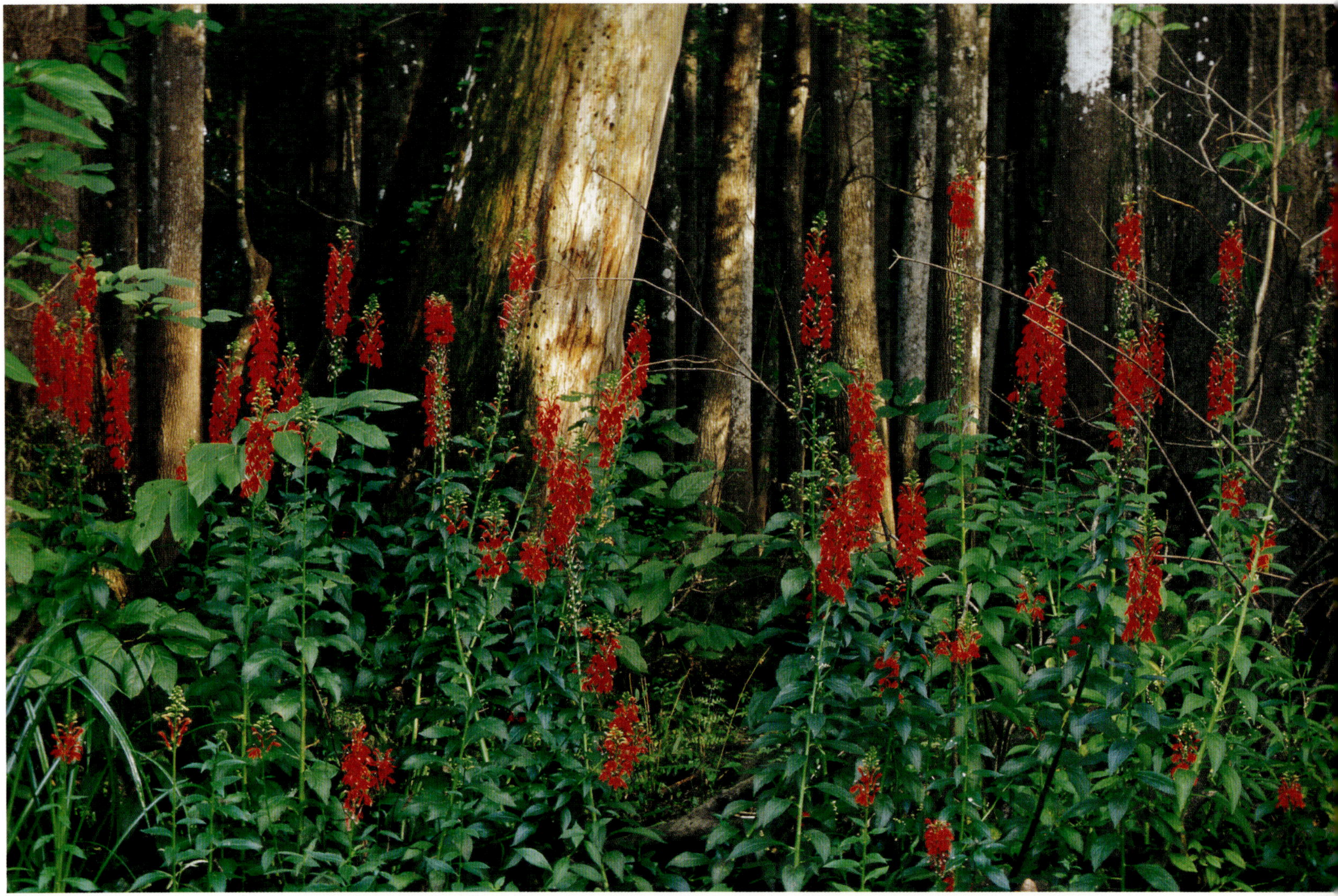

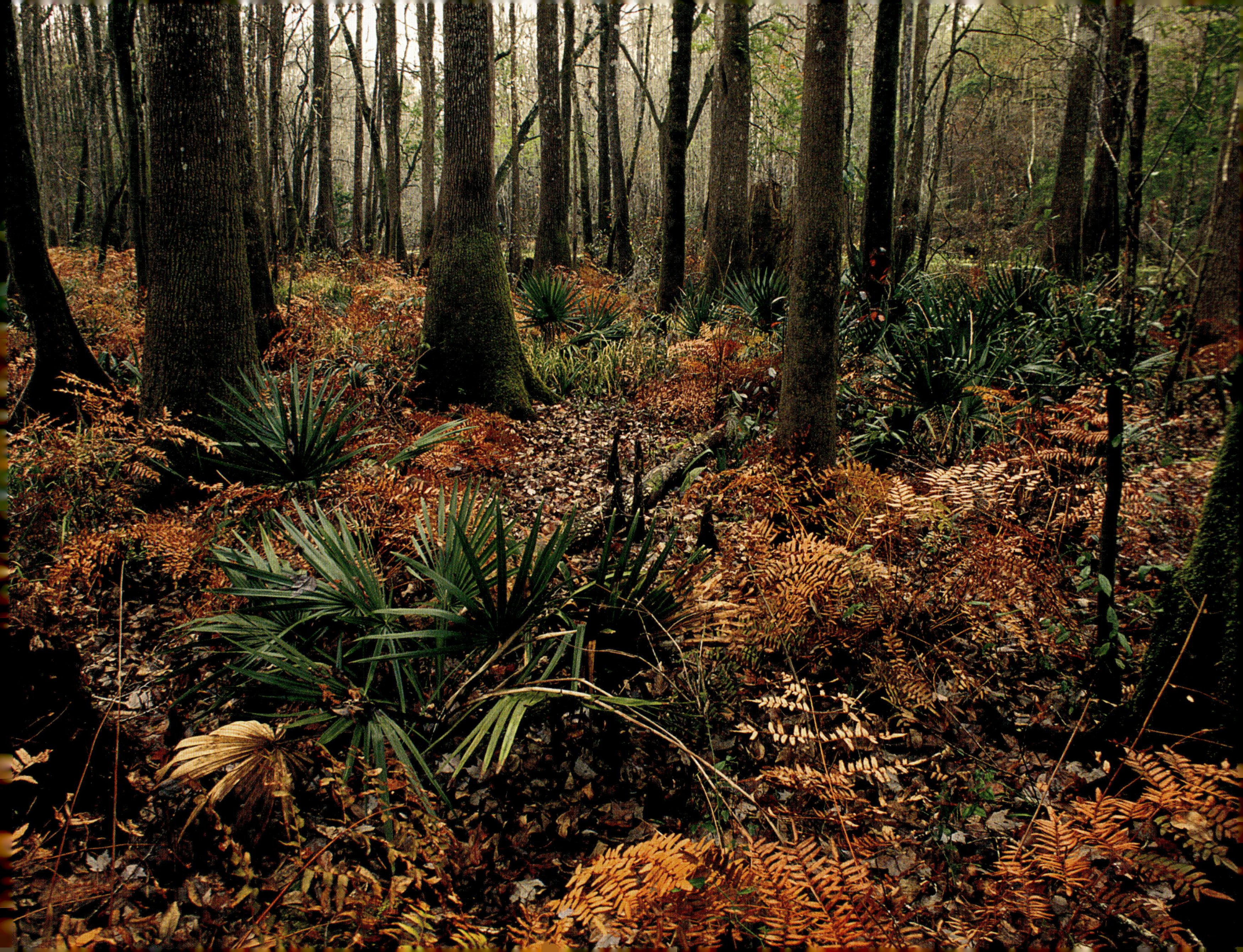

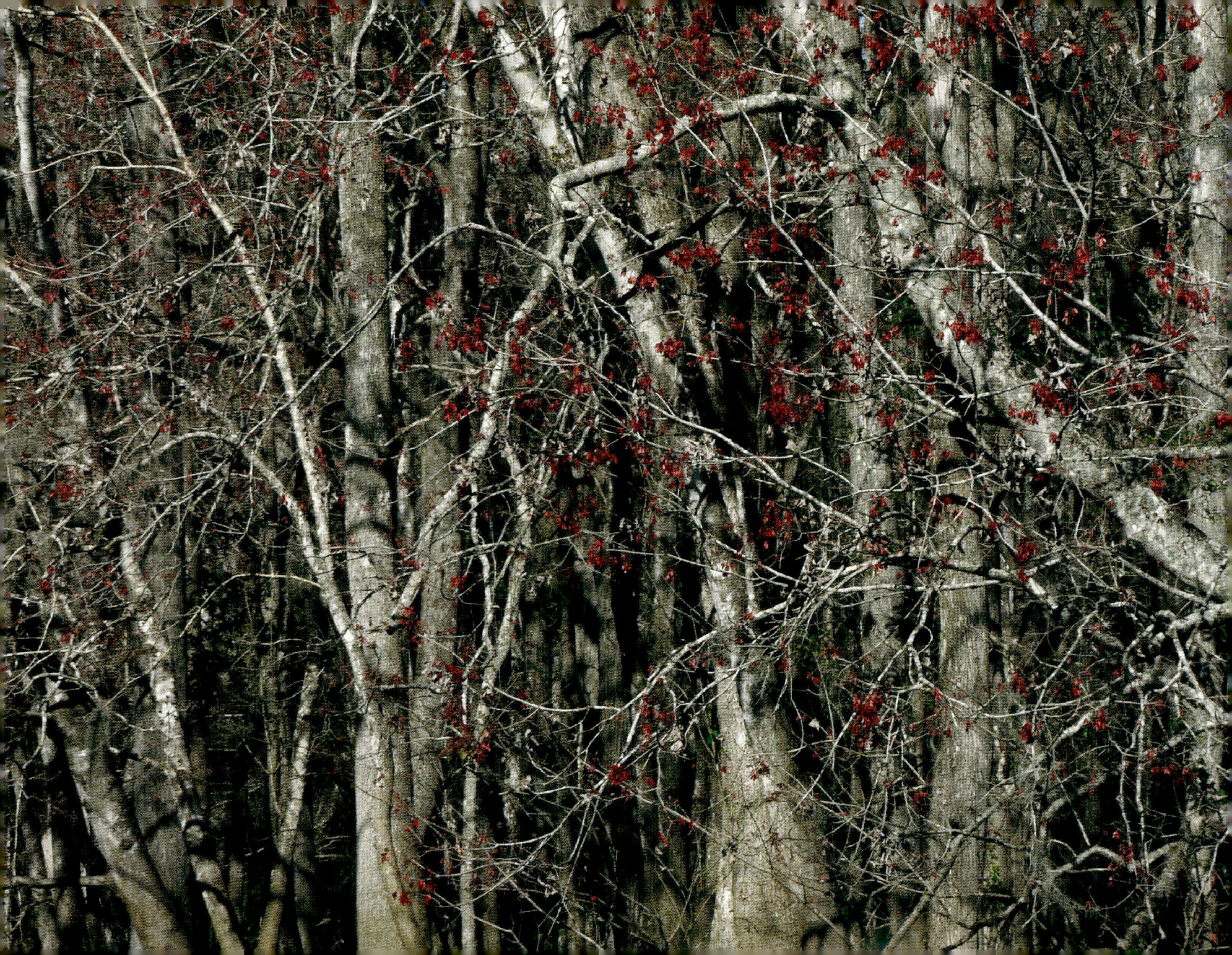

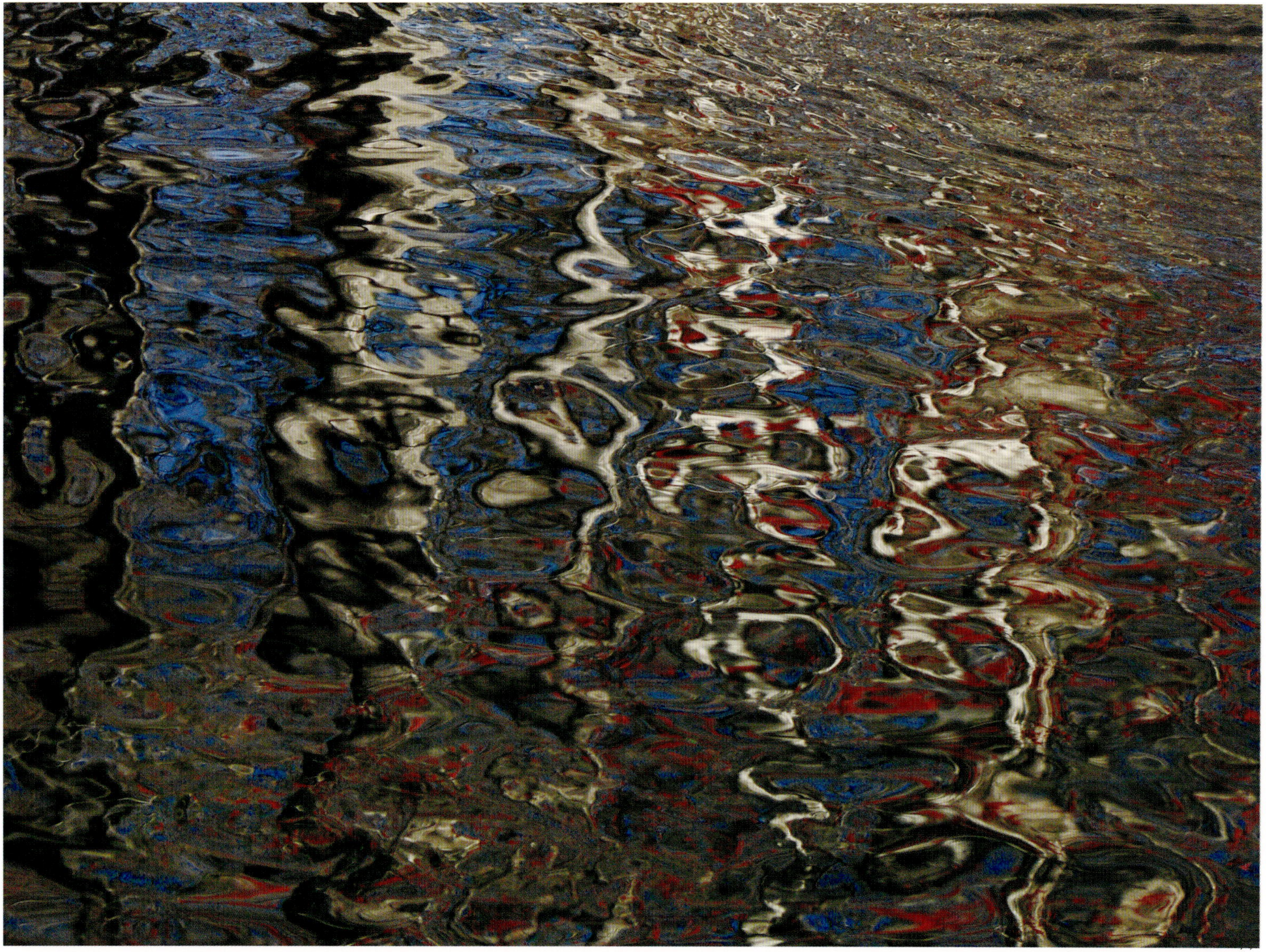

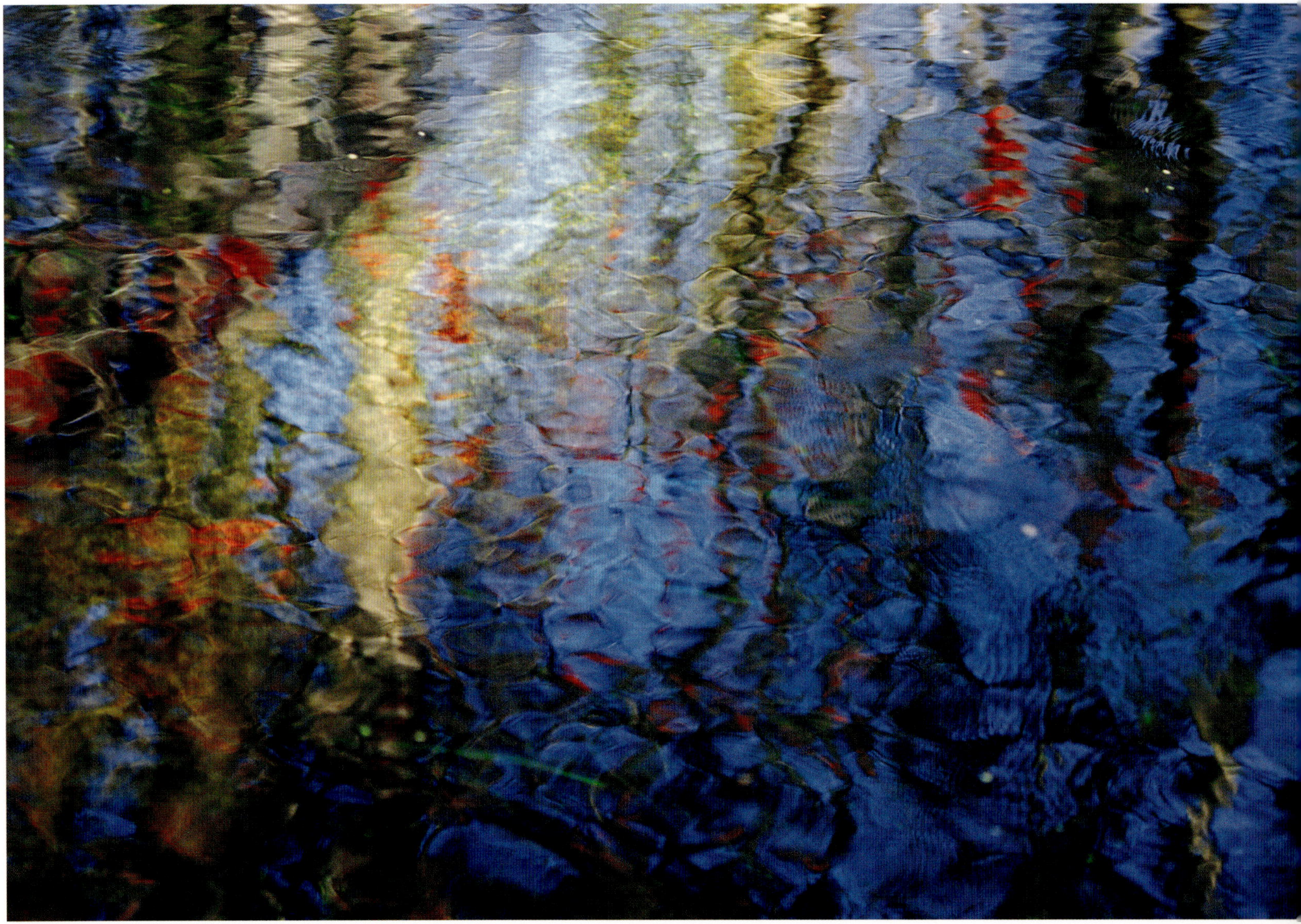

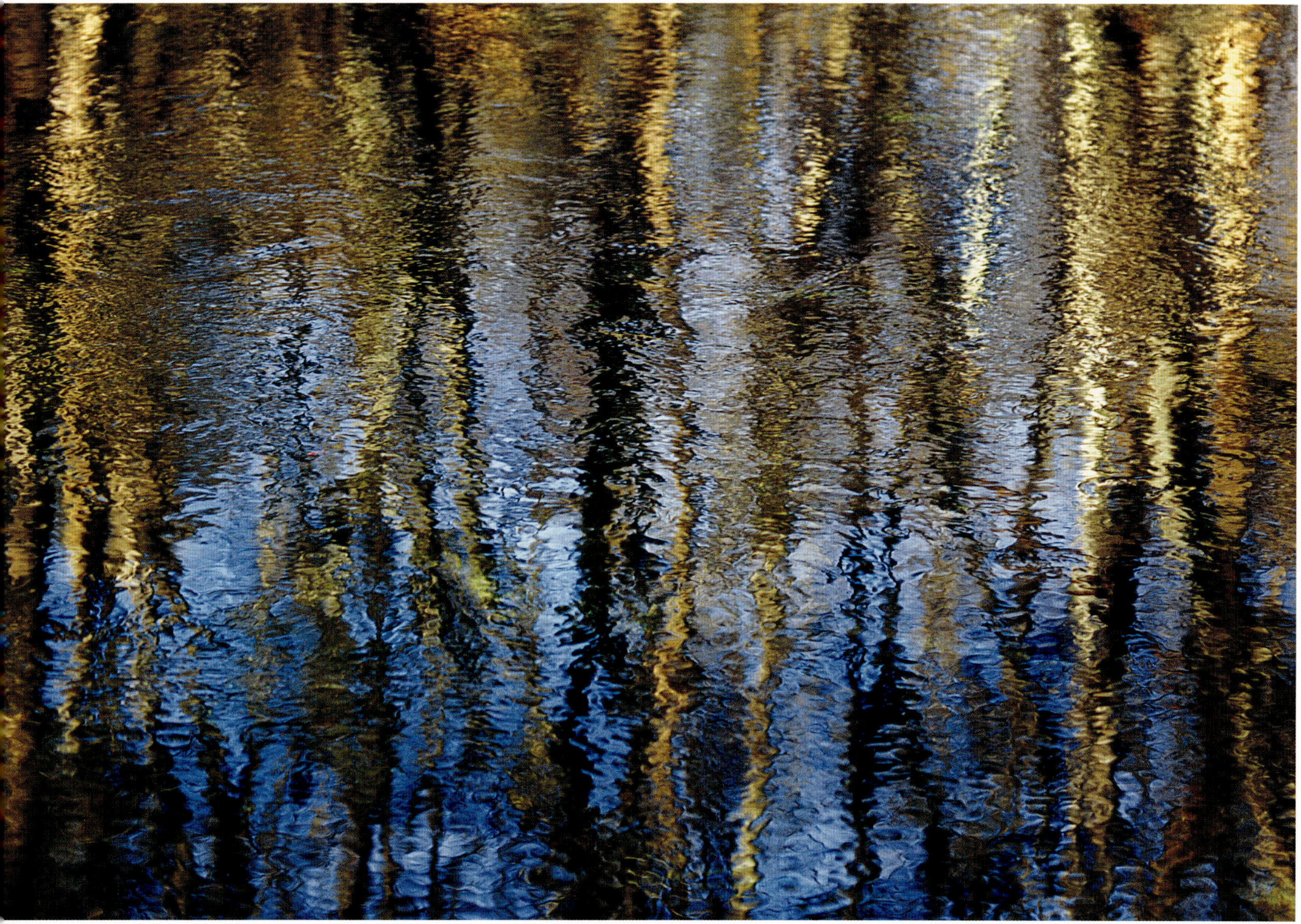

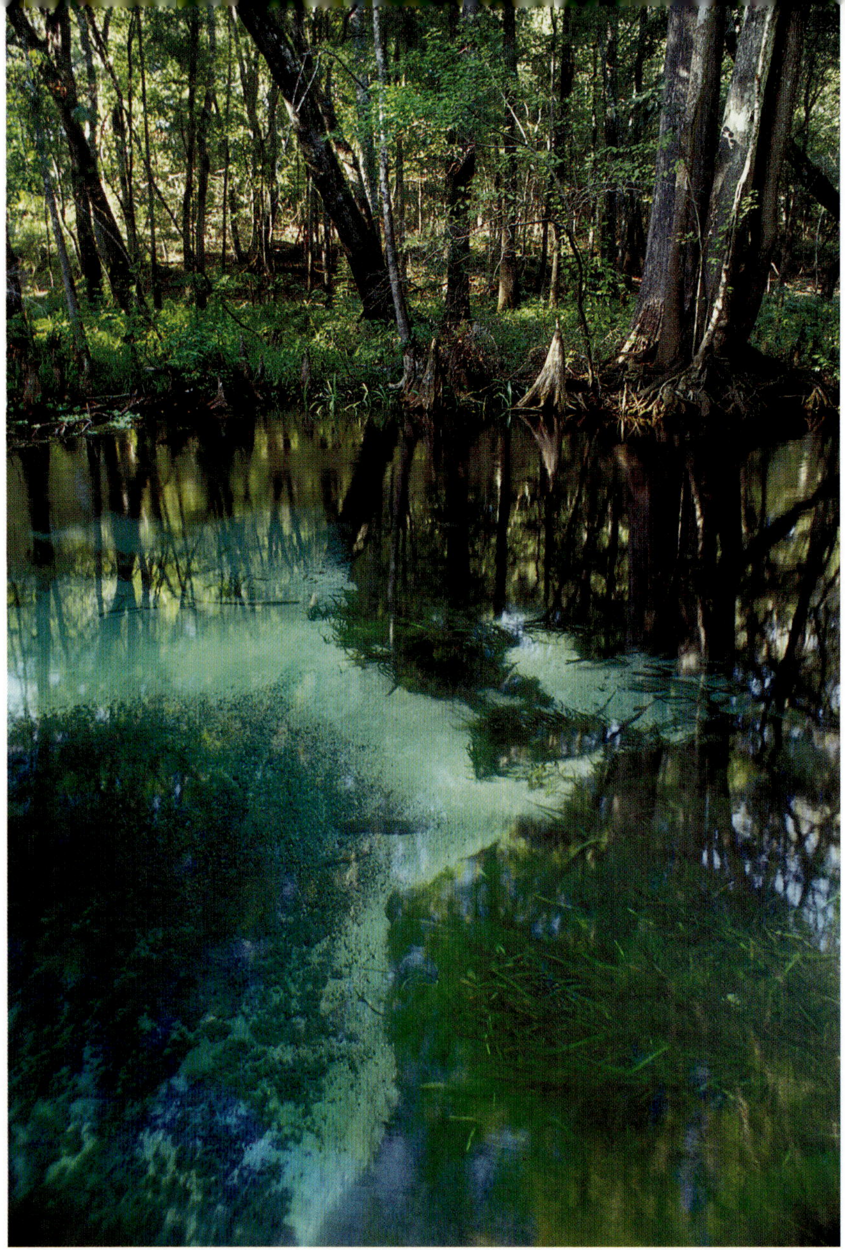

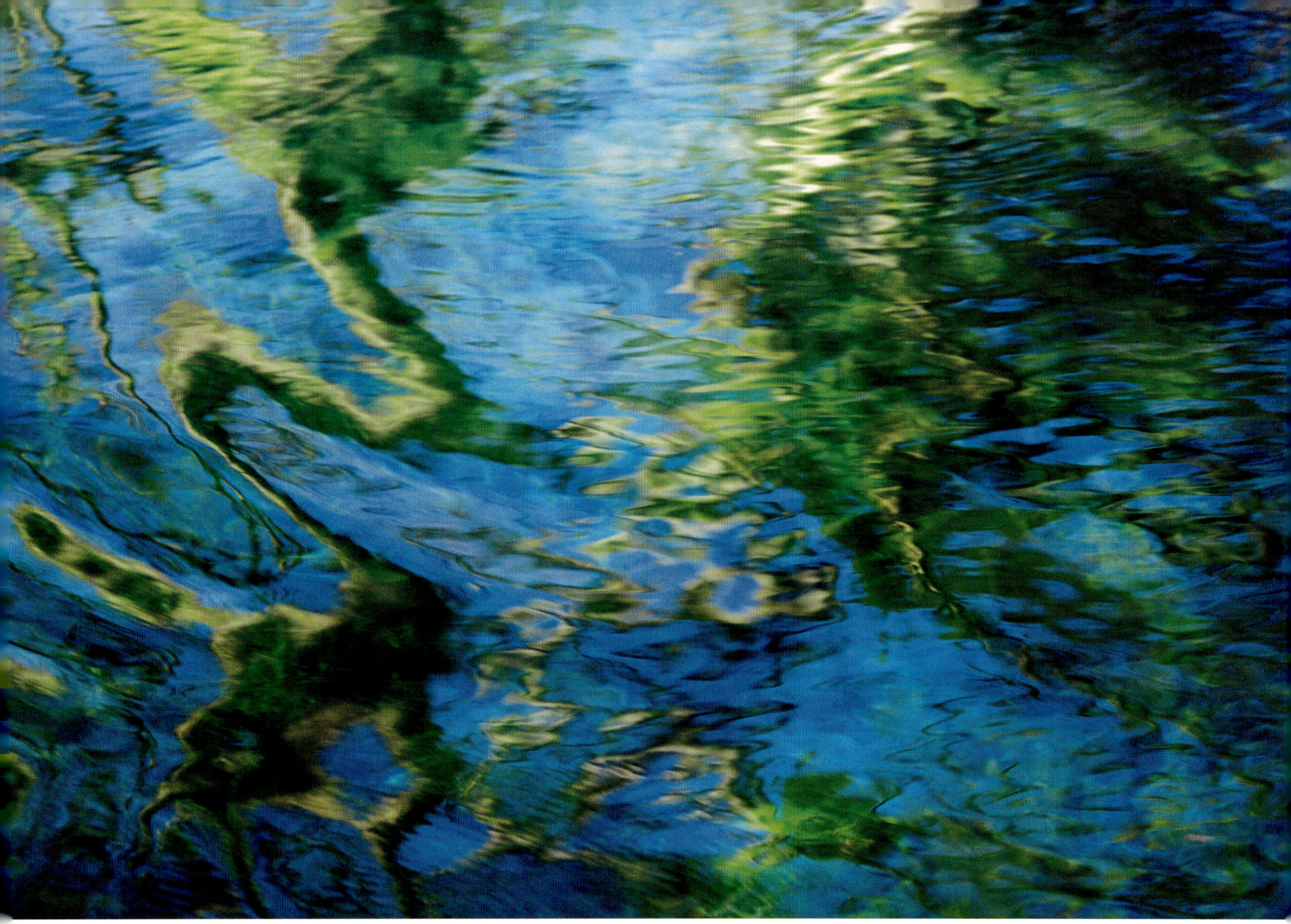

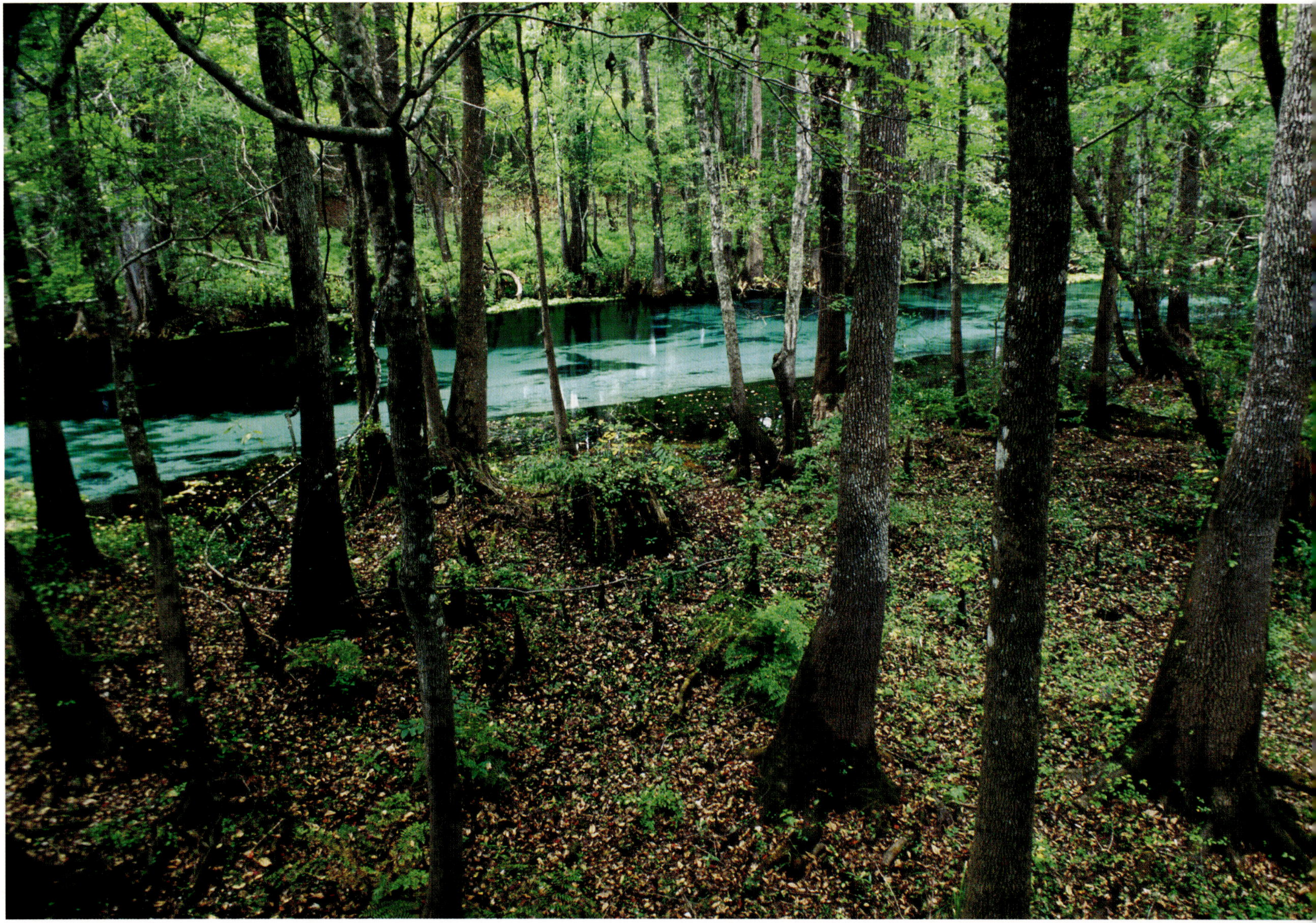

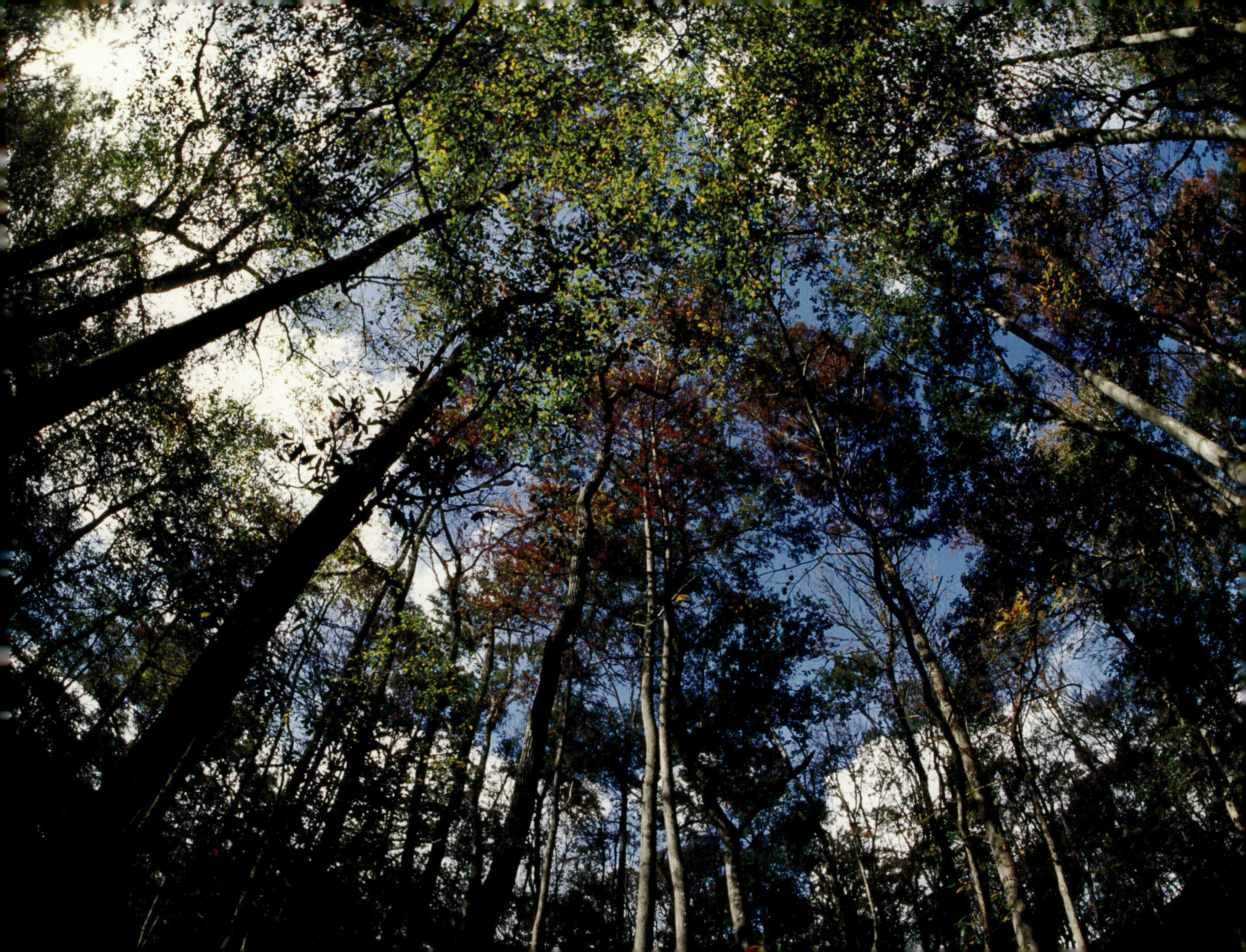

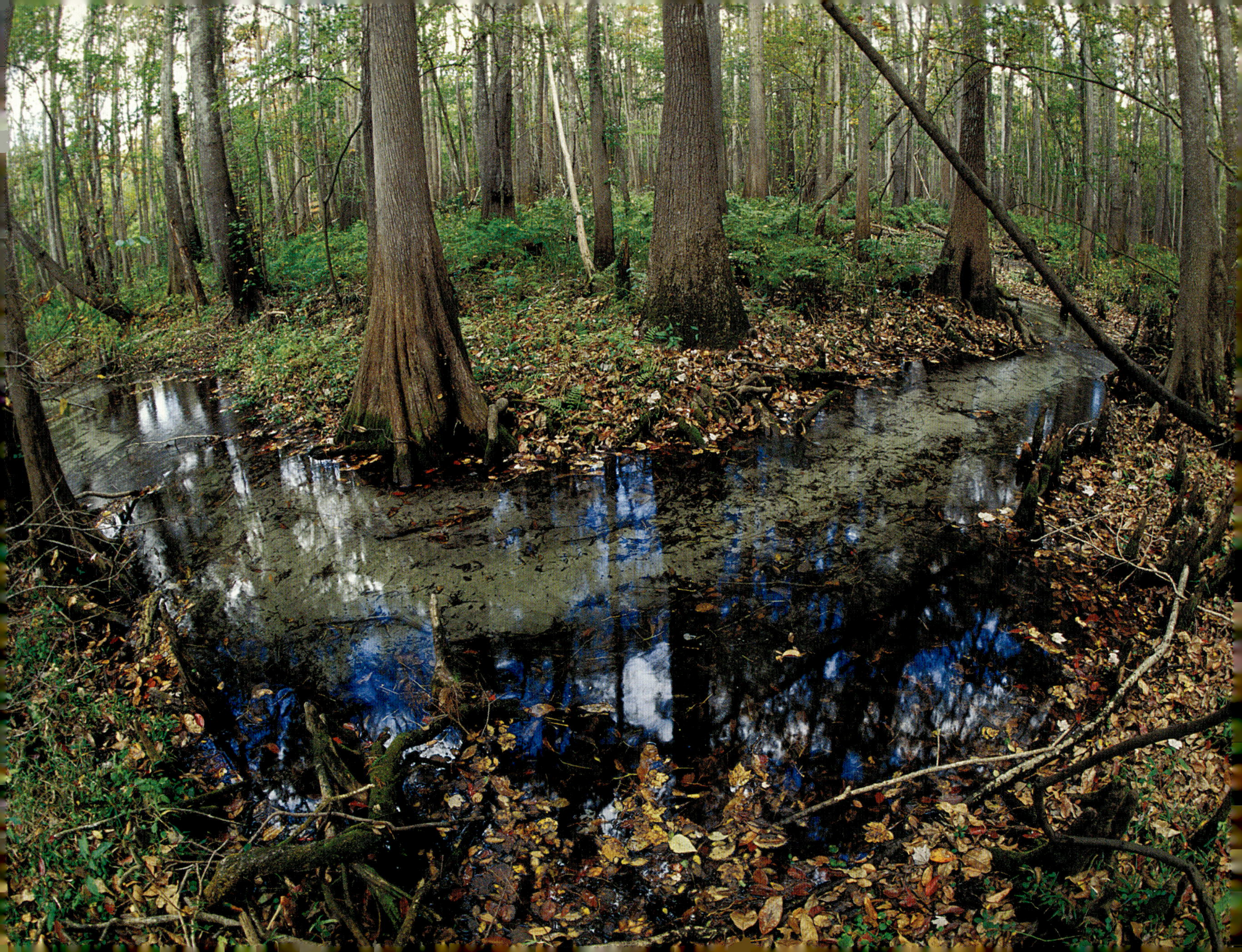

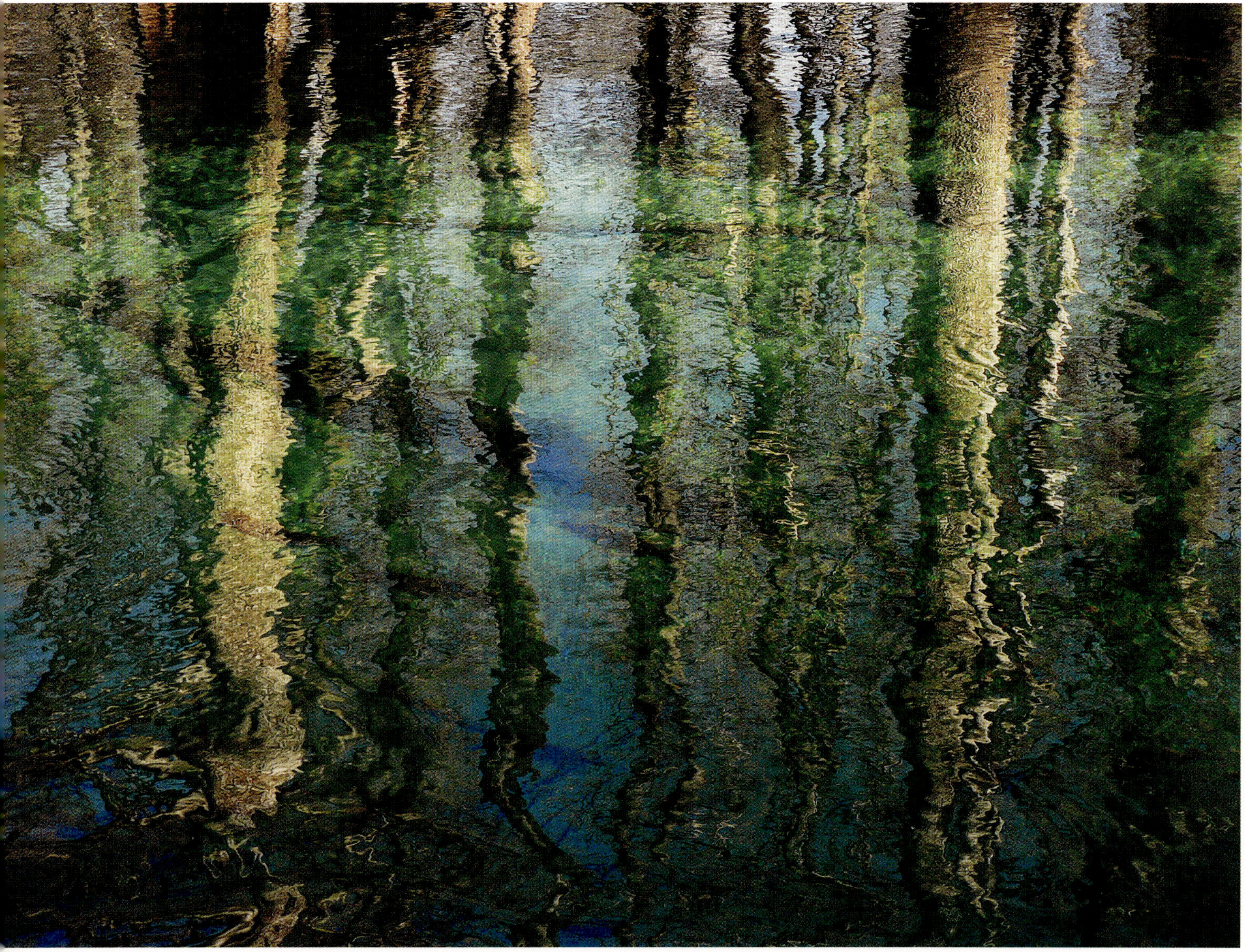

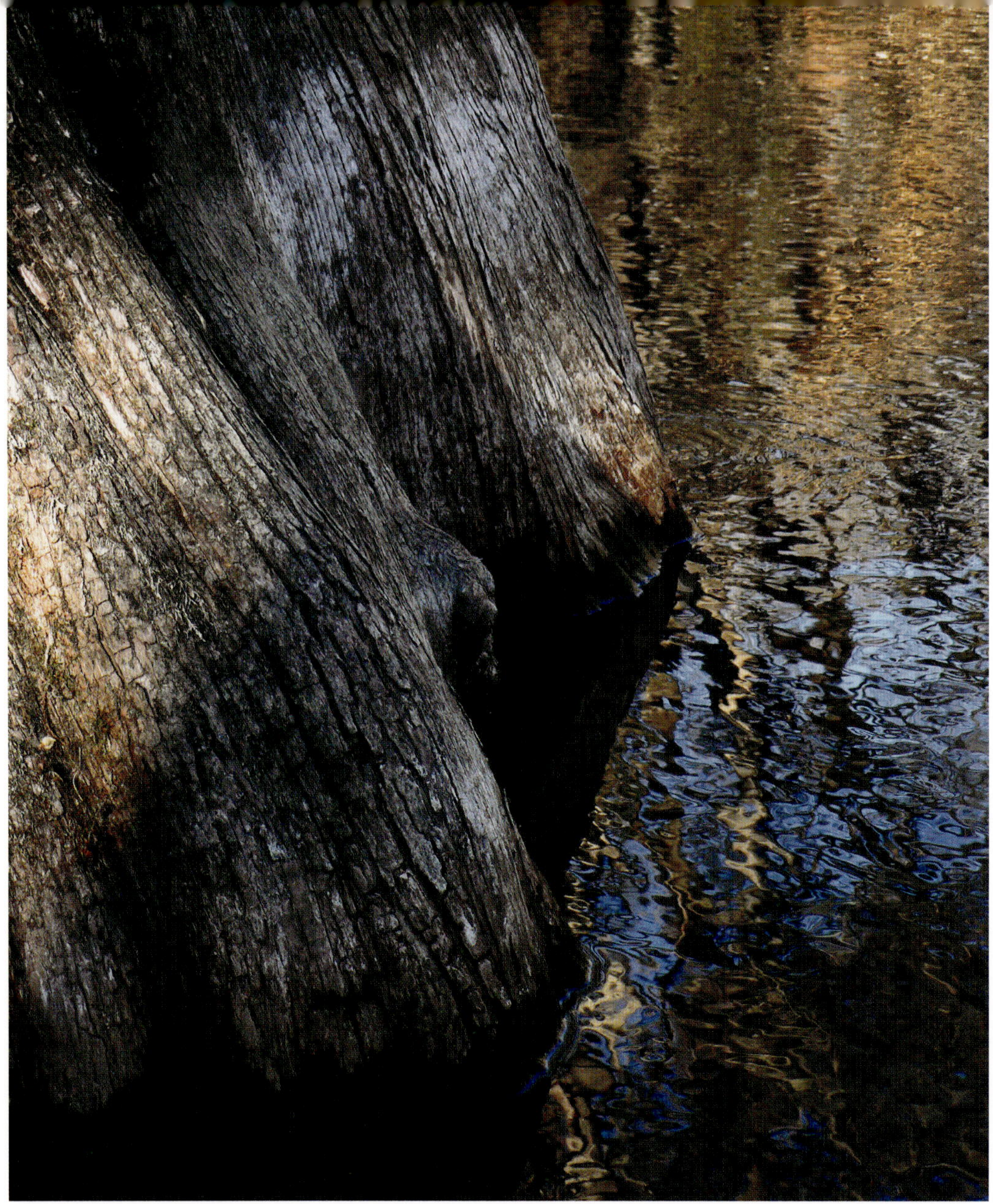

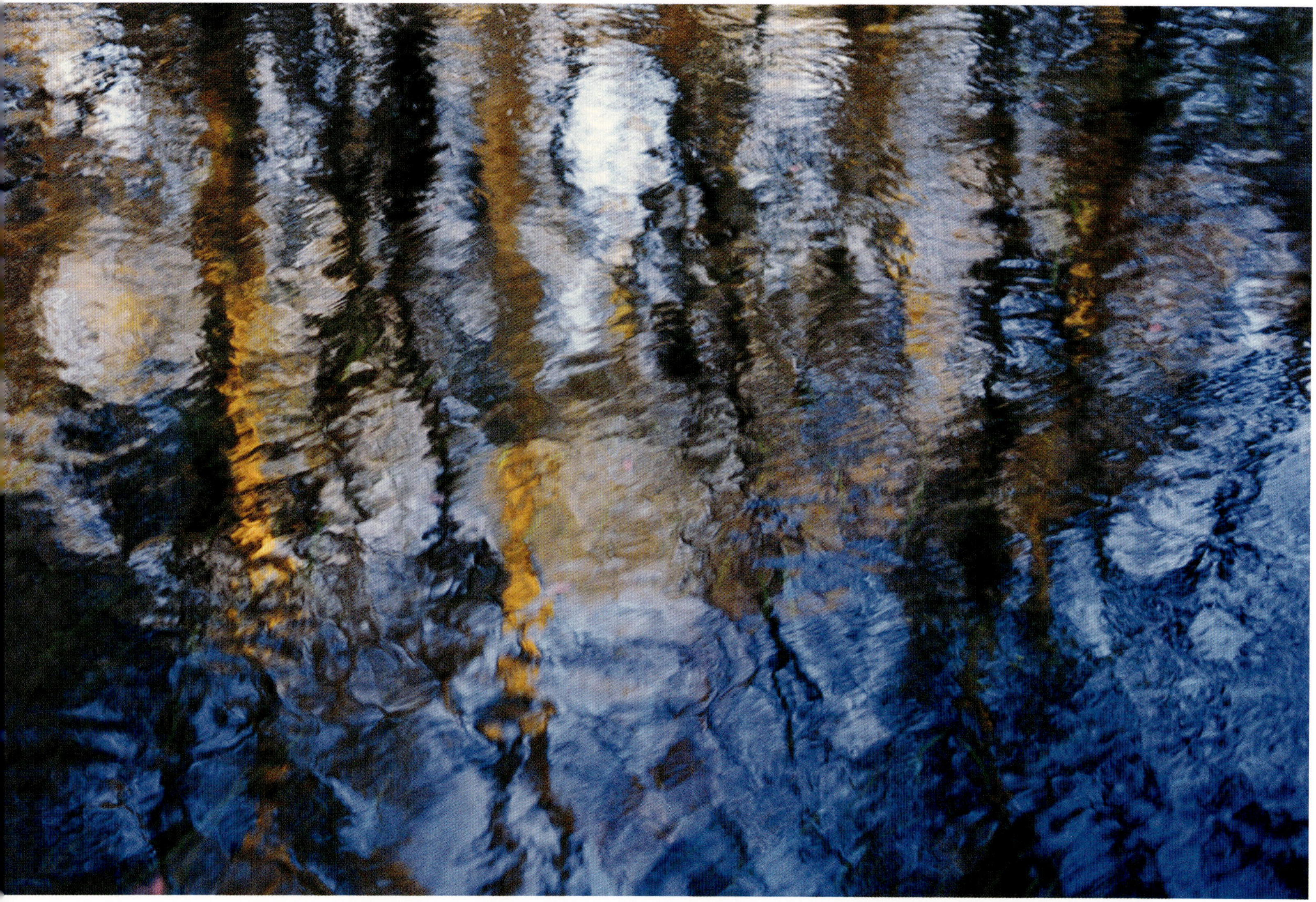

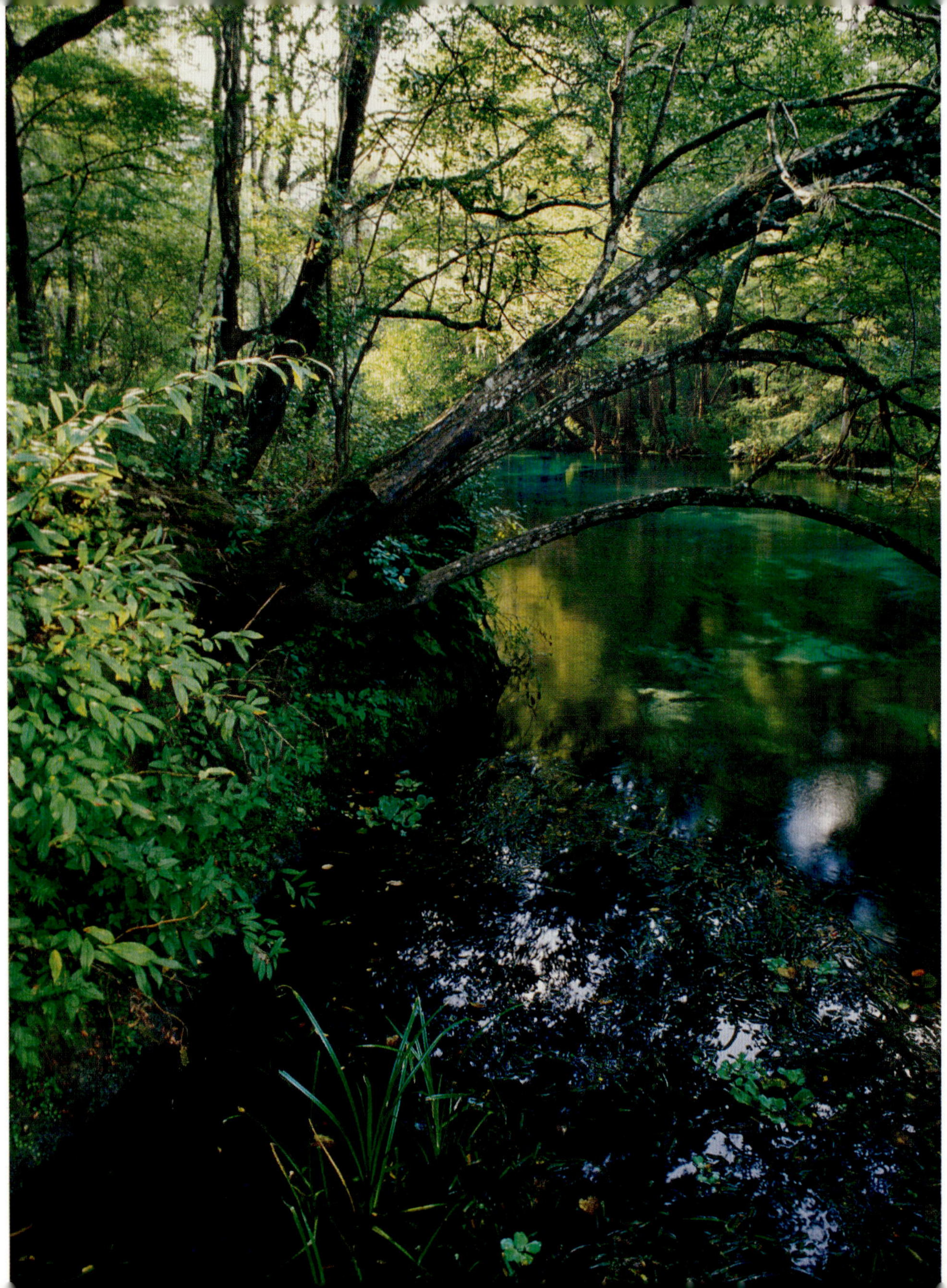

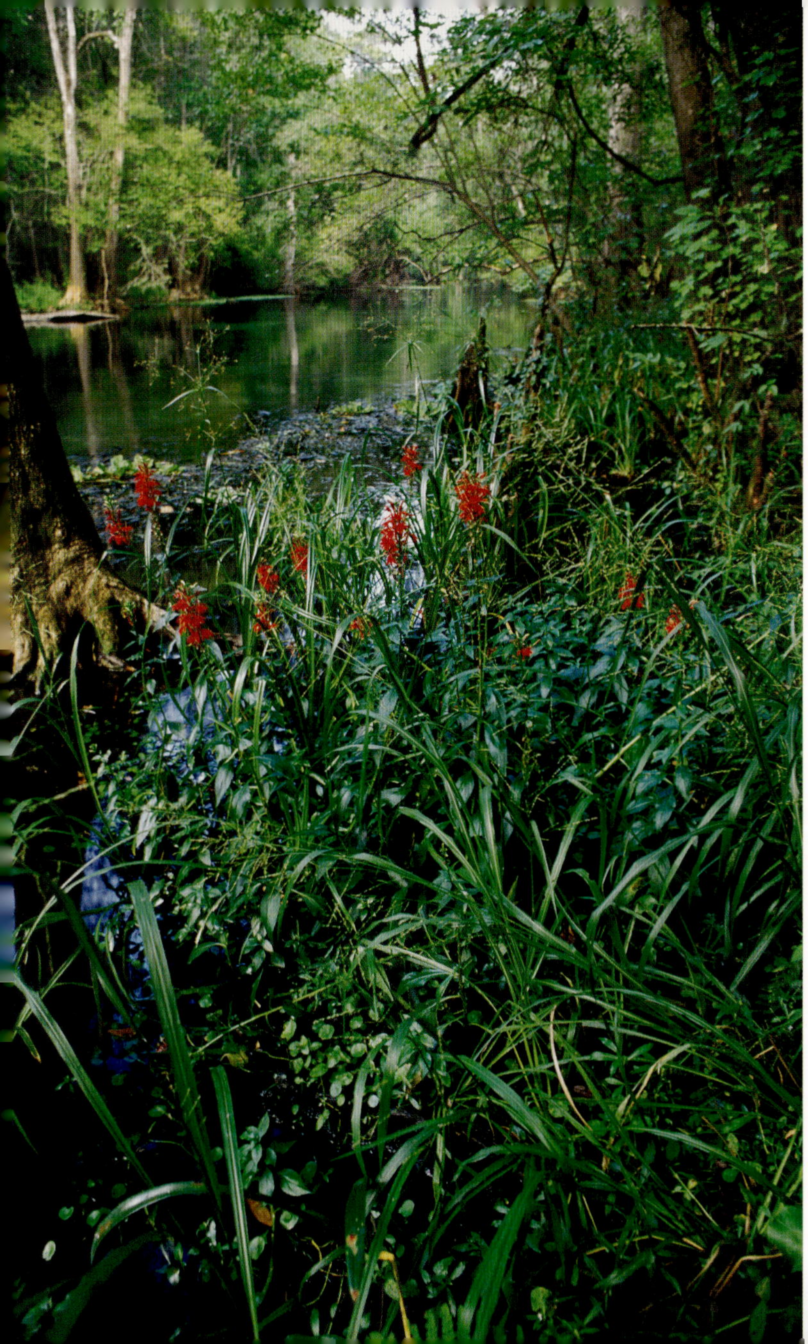

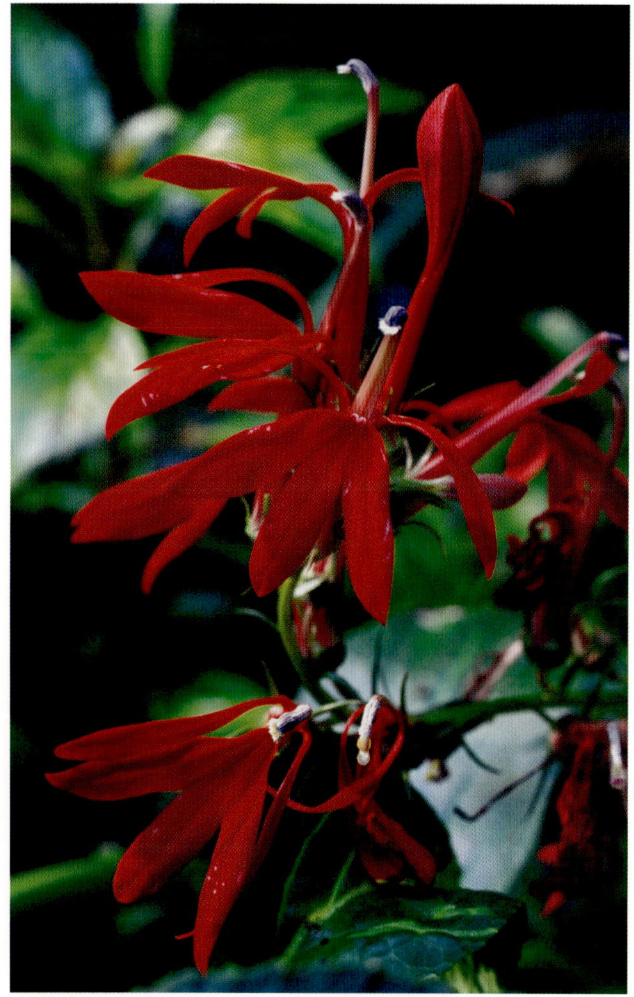

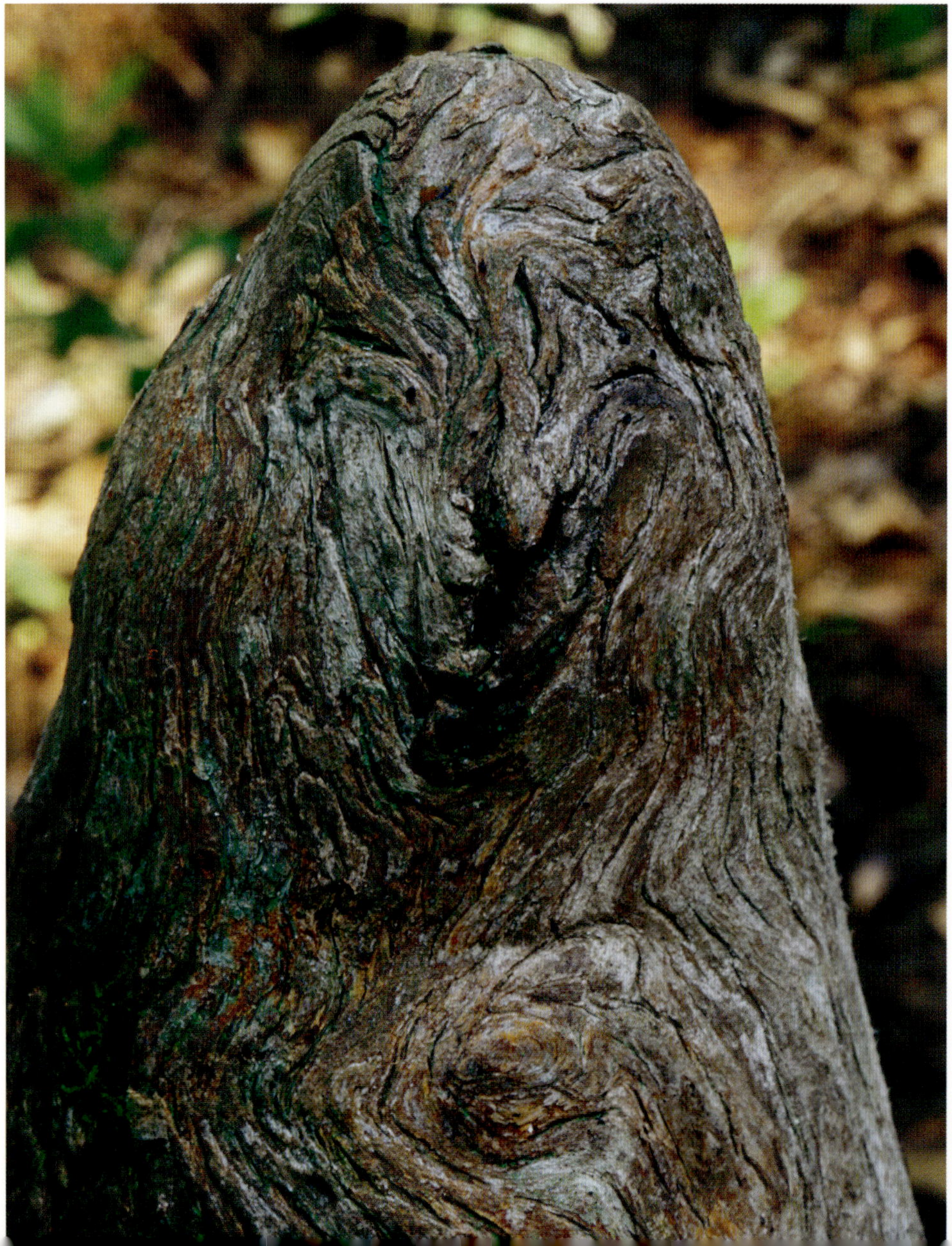

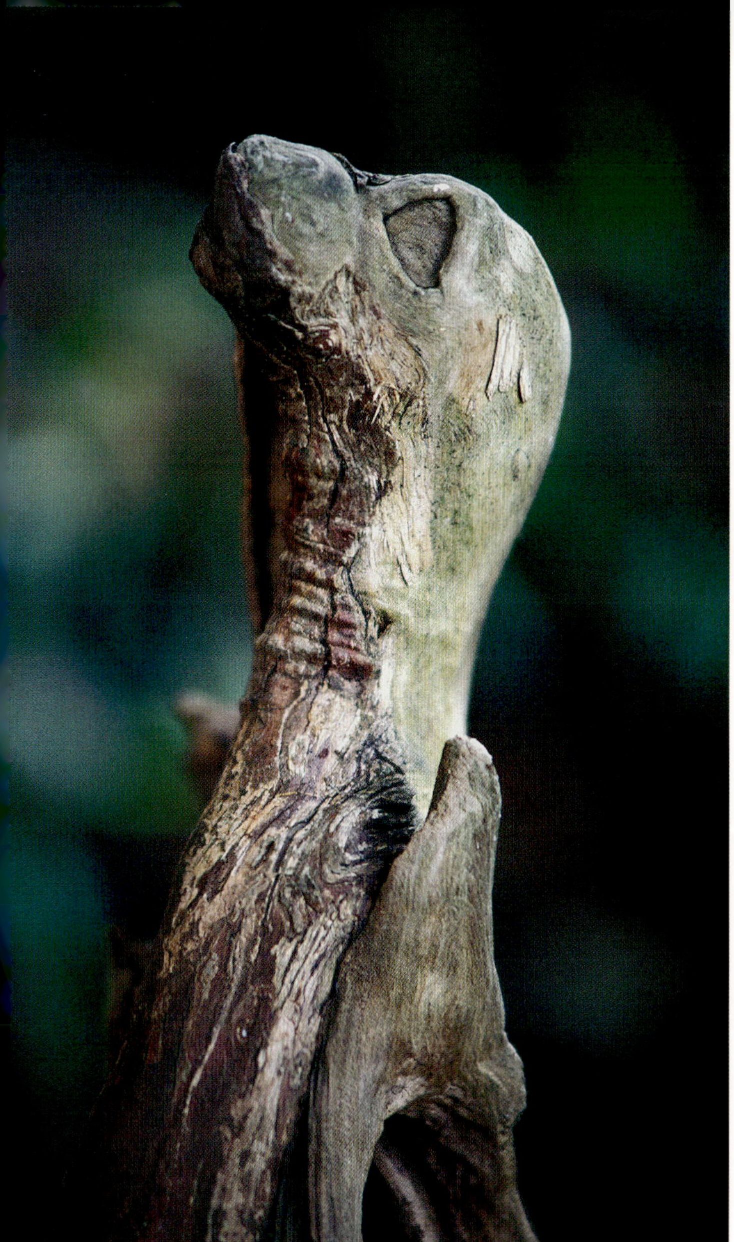

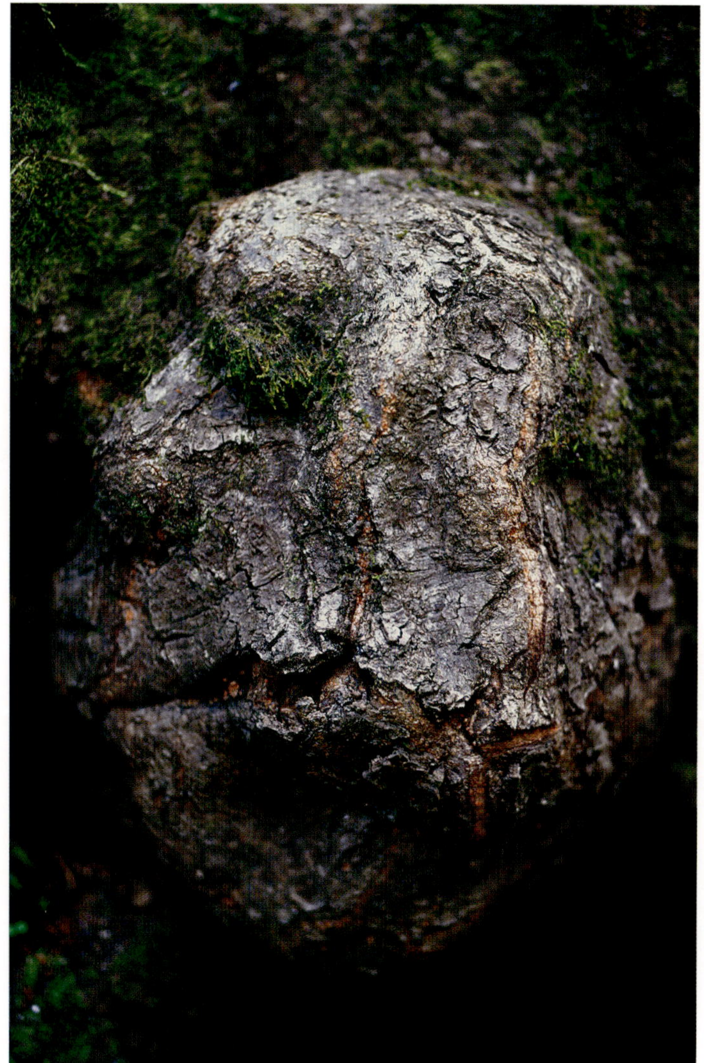

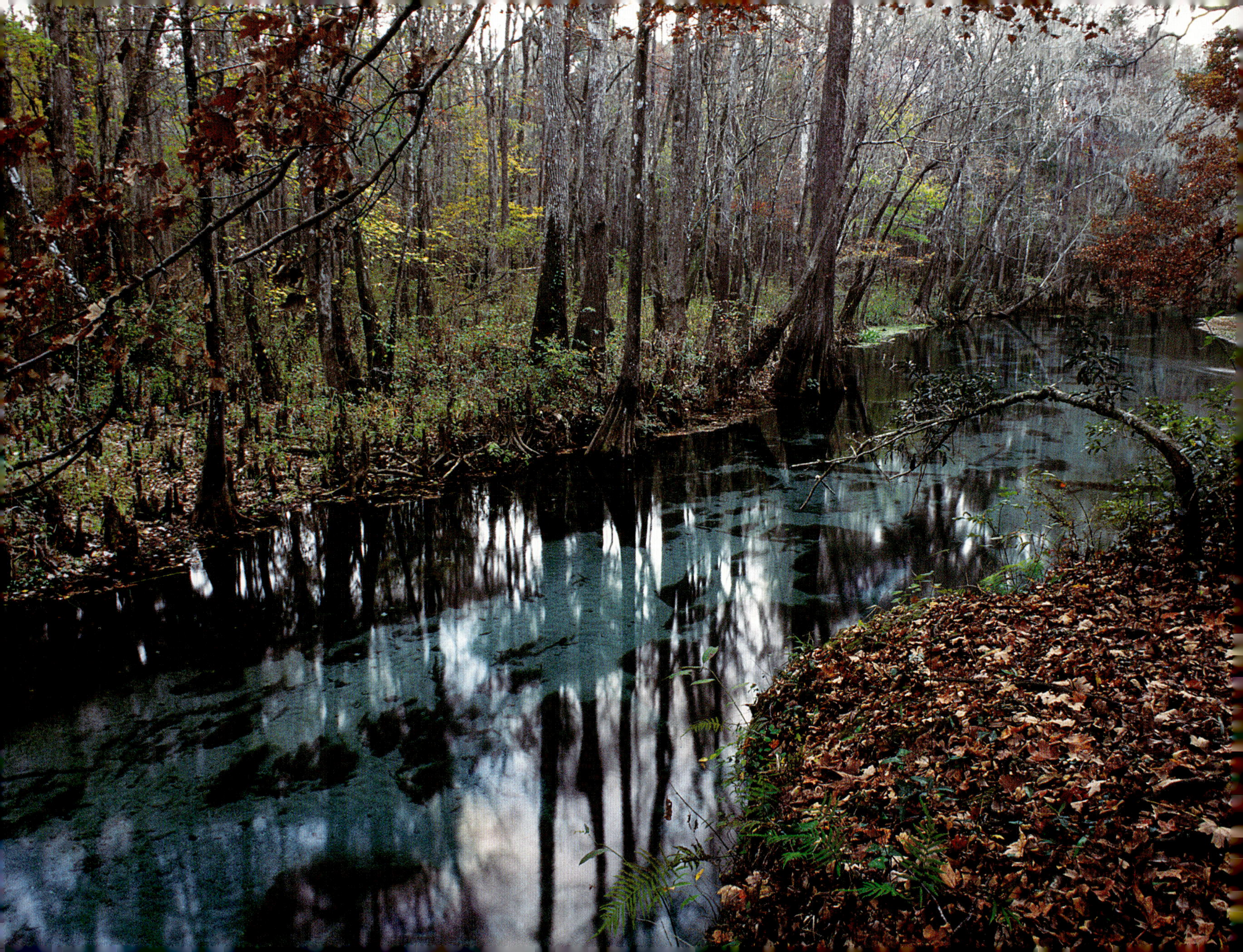

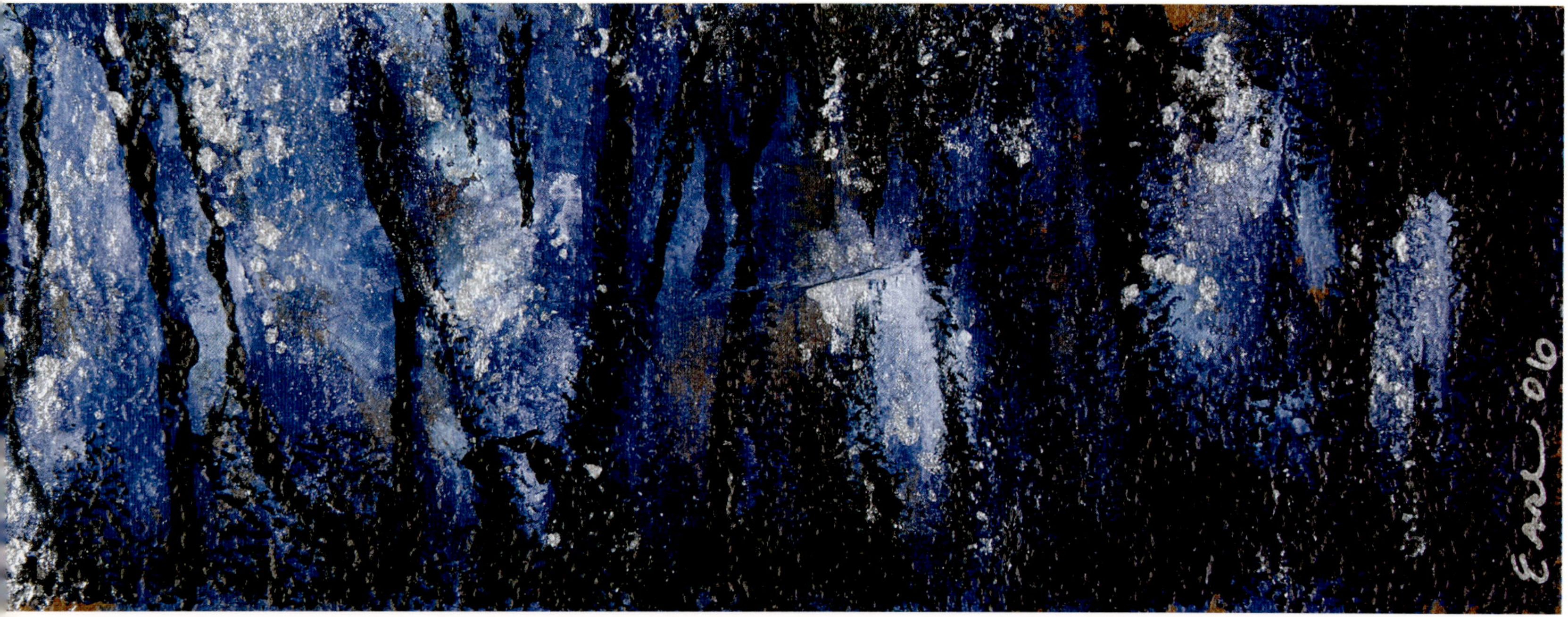

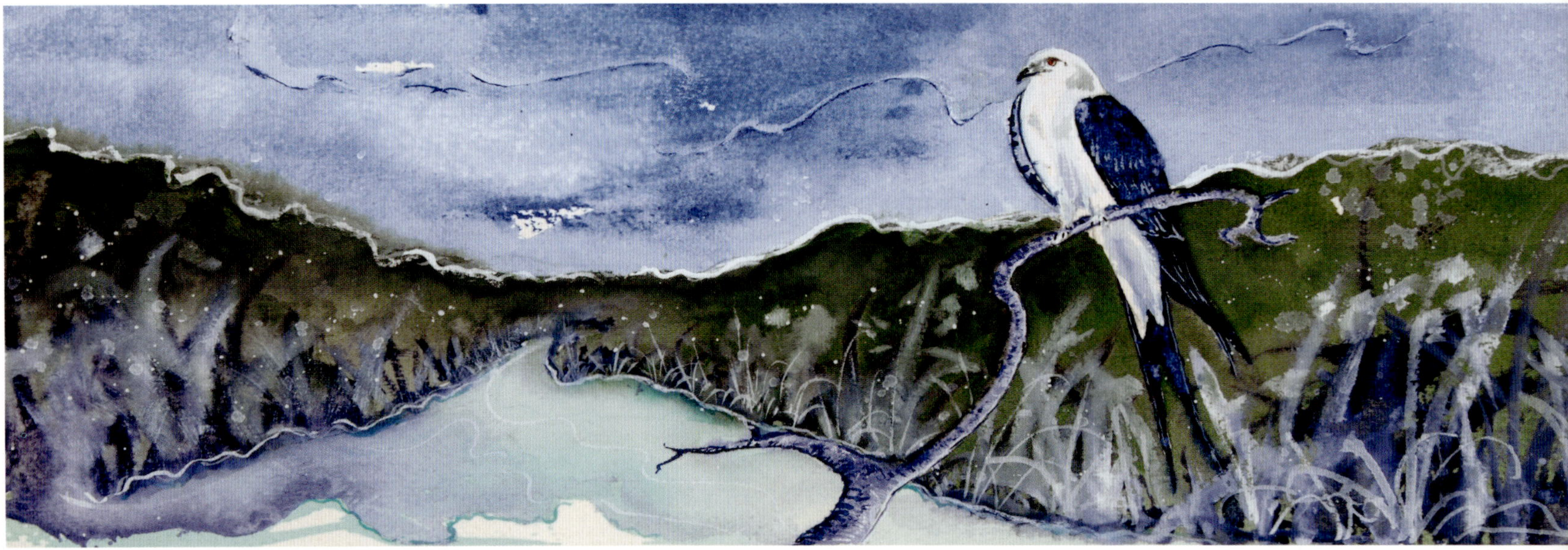

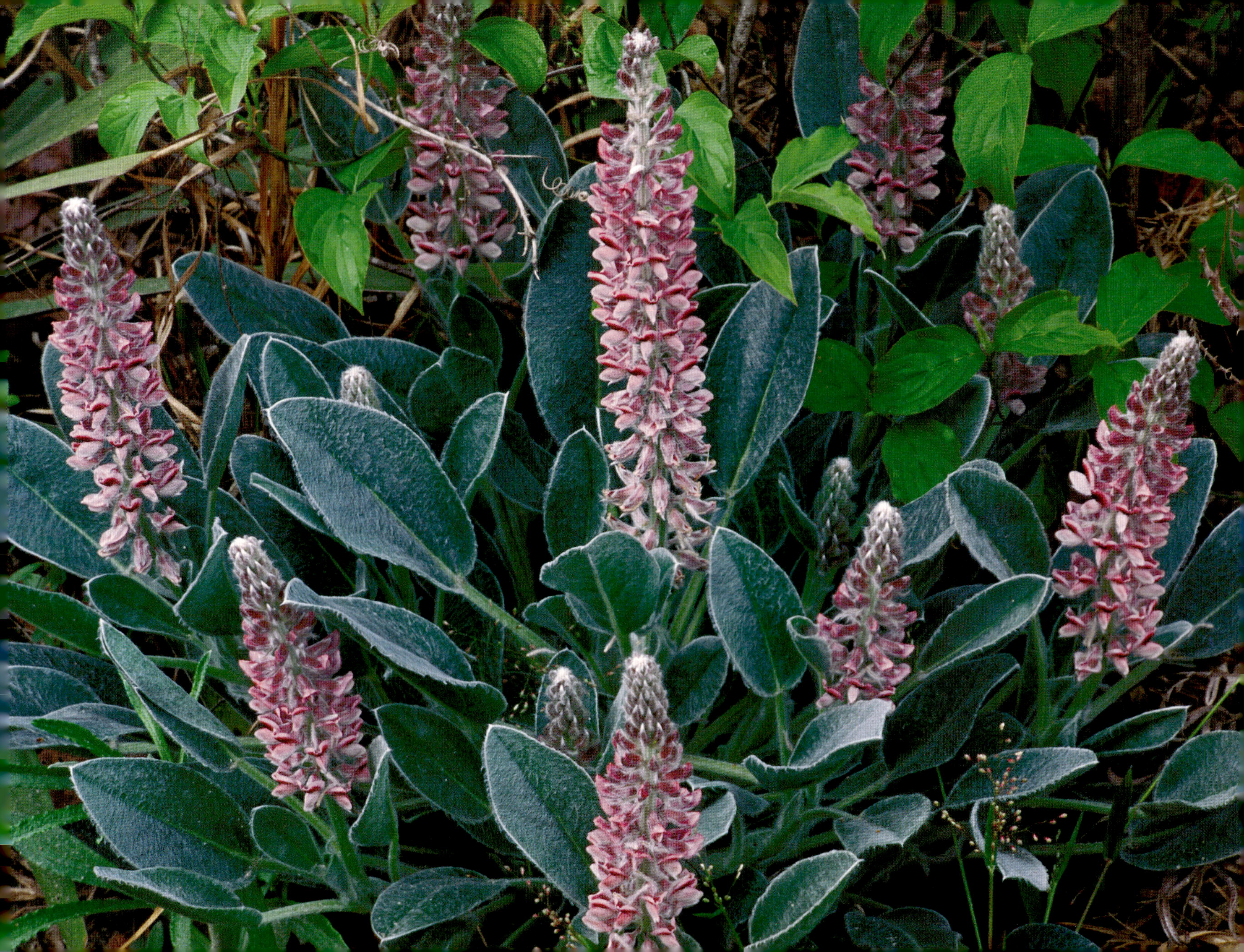

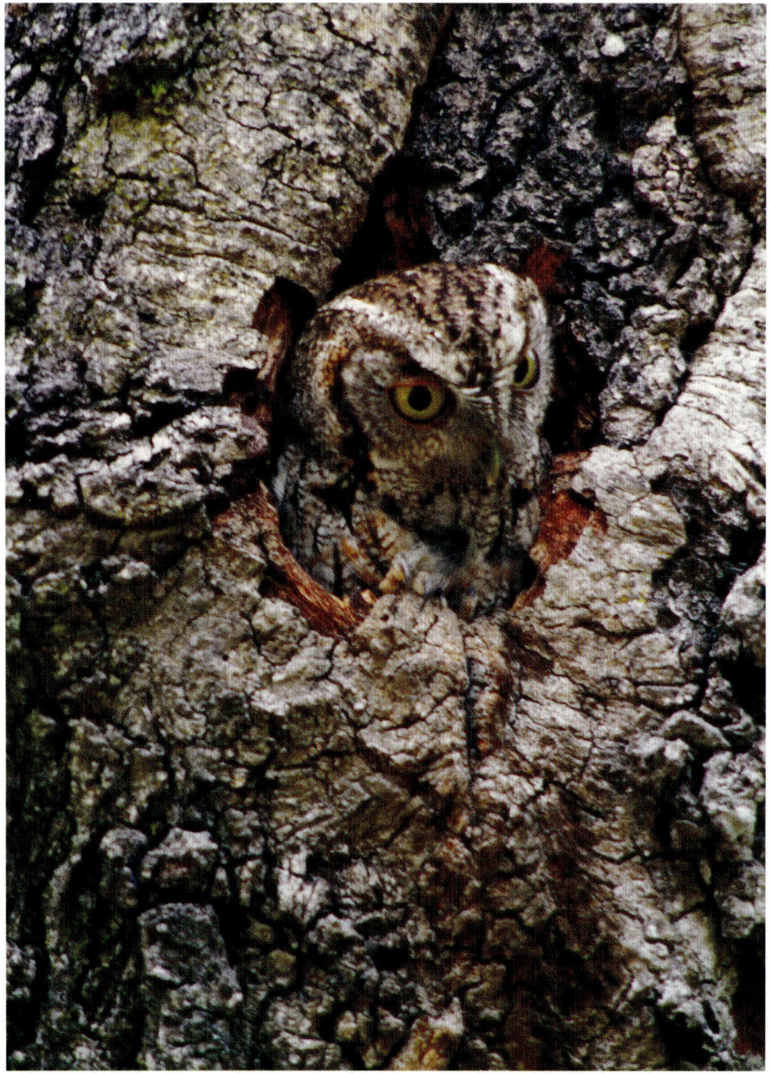

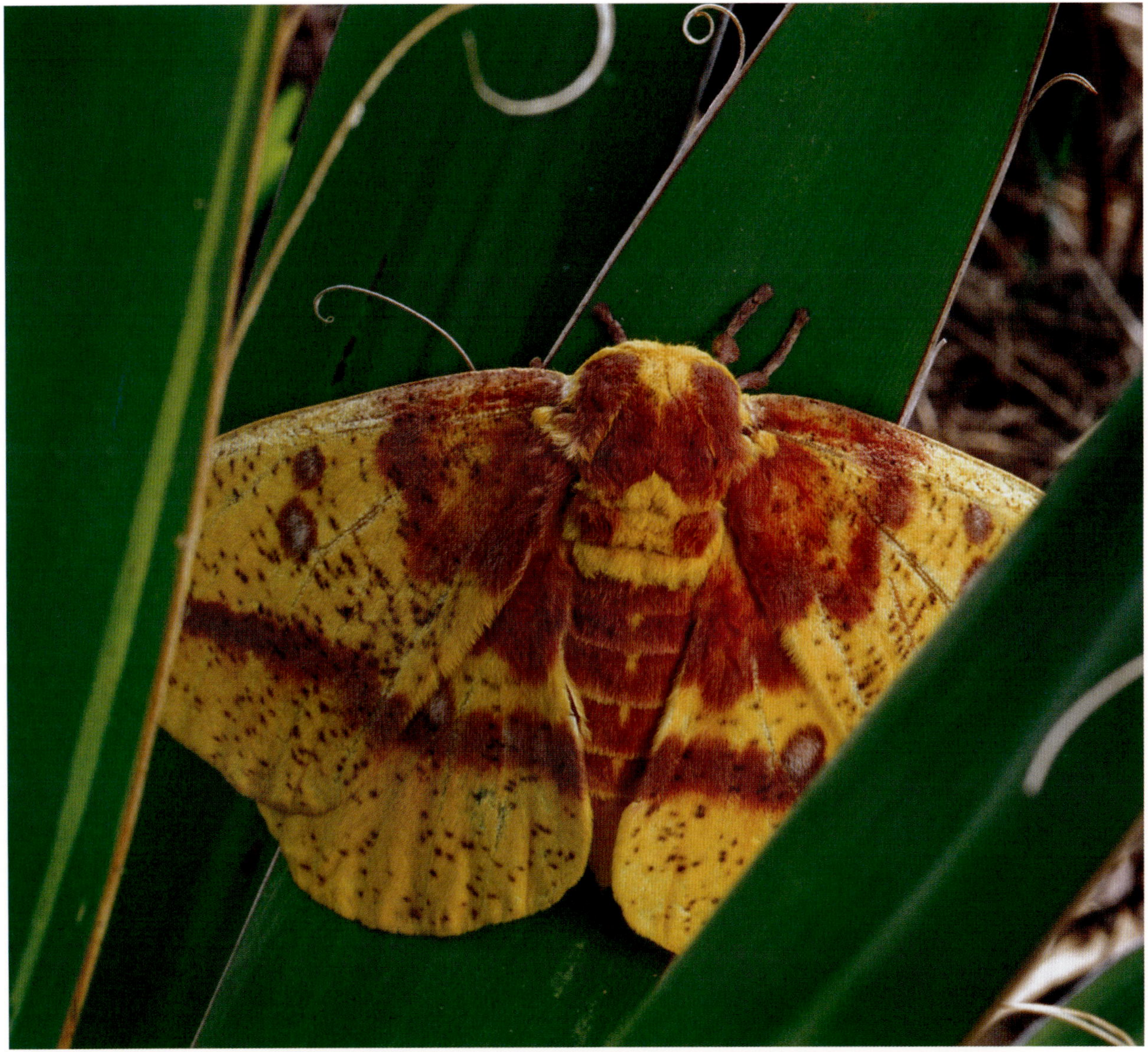

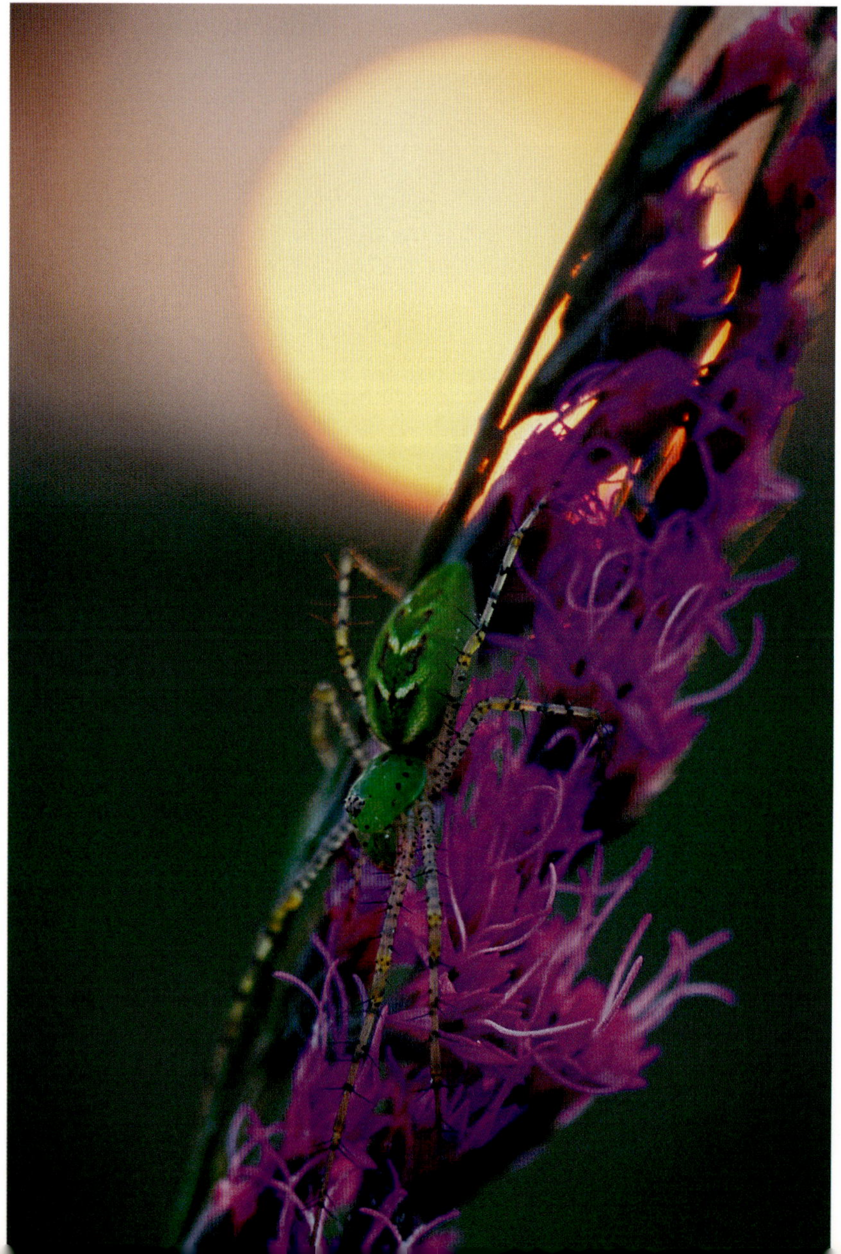

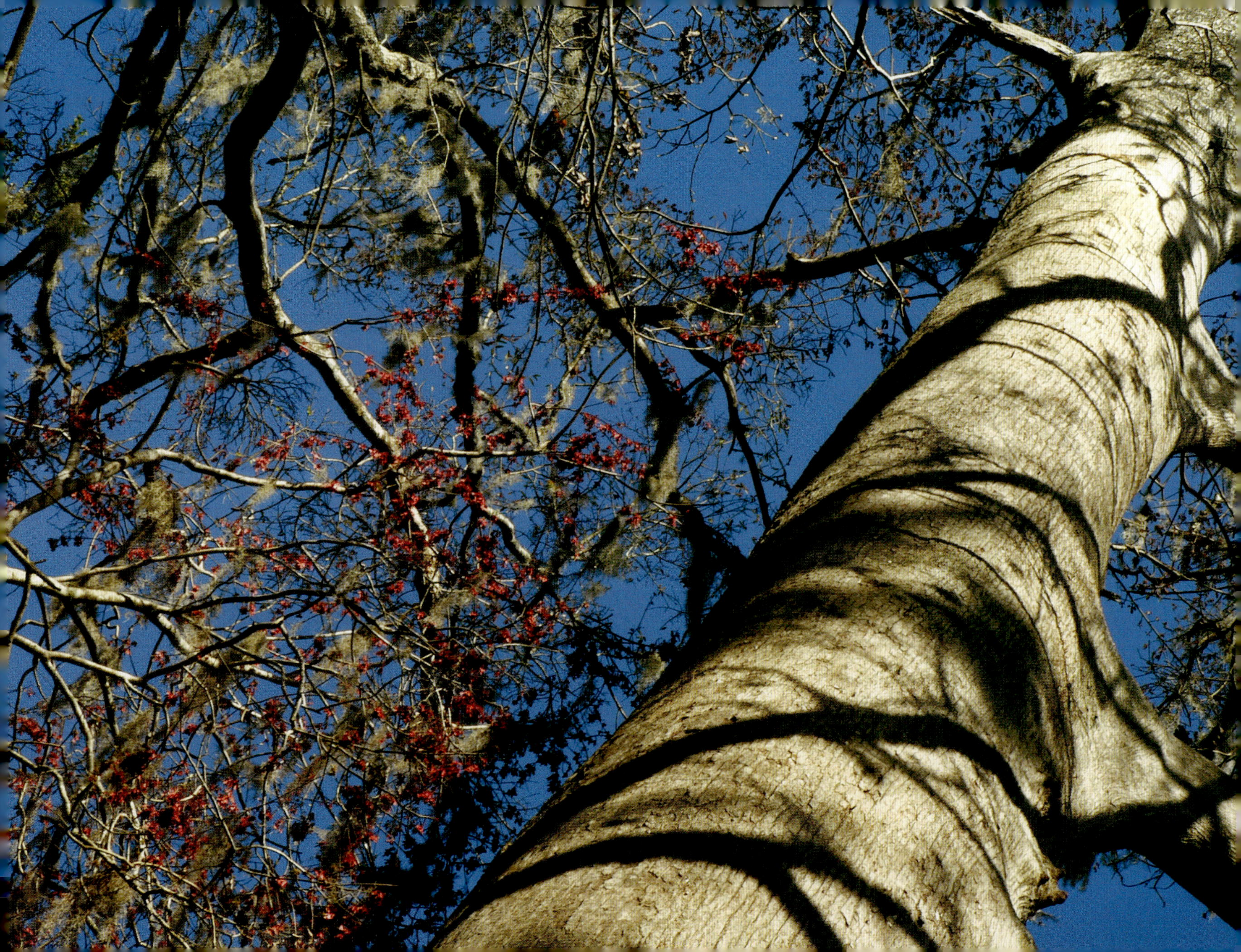

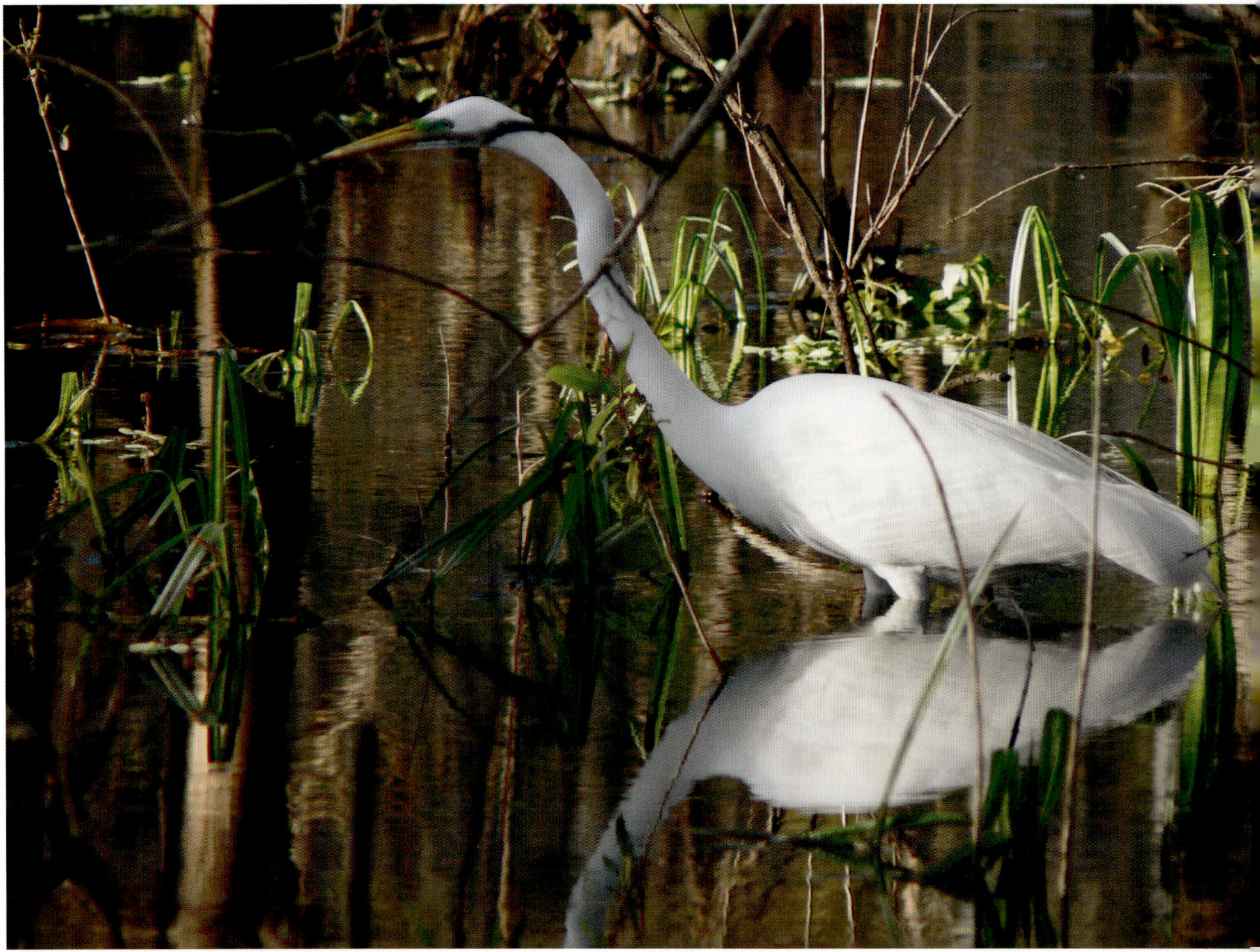

observing the observer
watching the watchful
stepping into the eyes of the great egret
a shift in focus, perspective, purpose
becoming one inside a feathered body
standing in the waters
fully intent

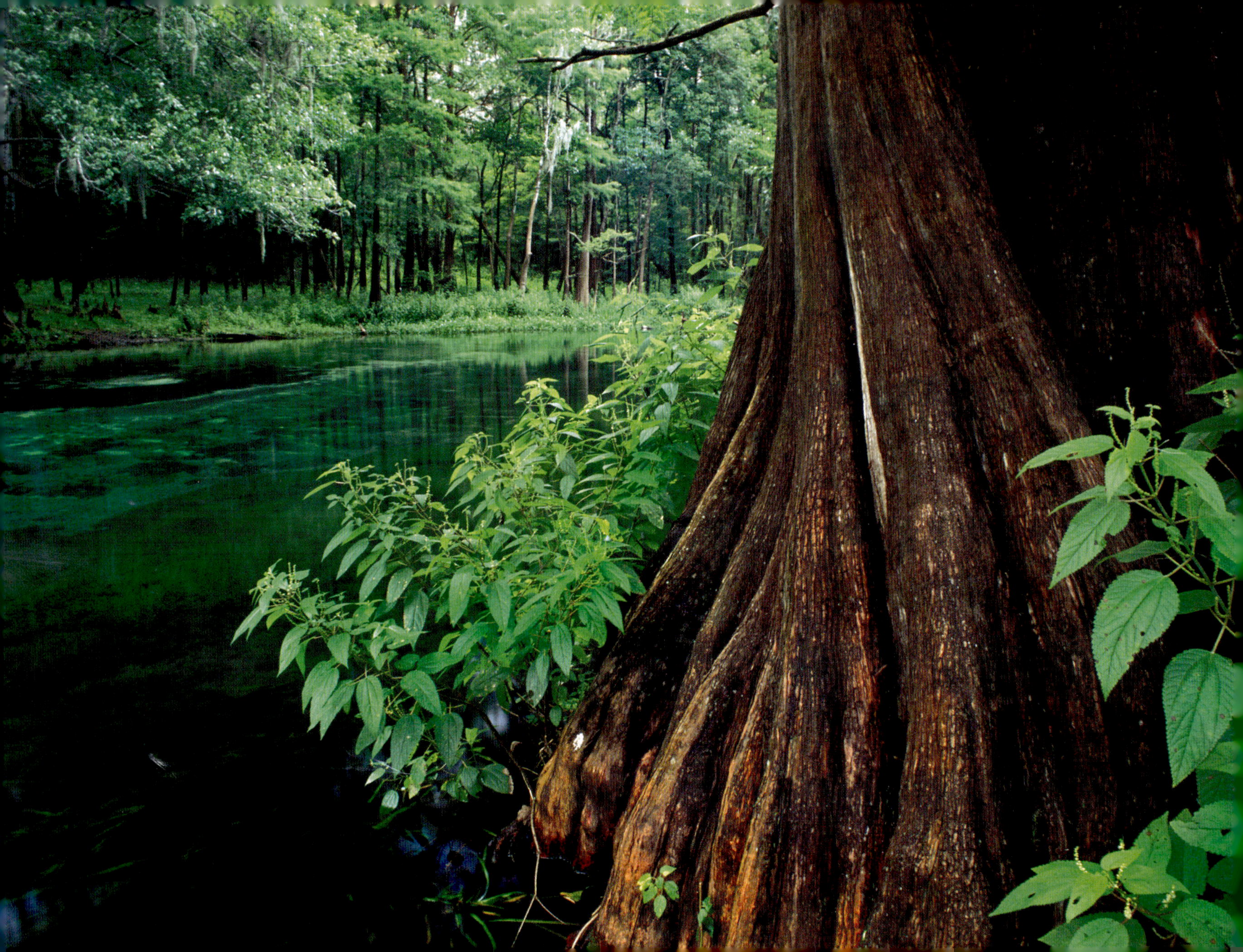

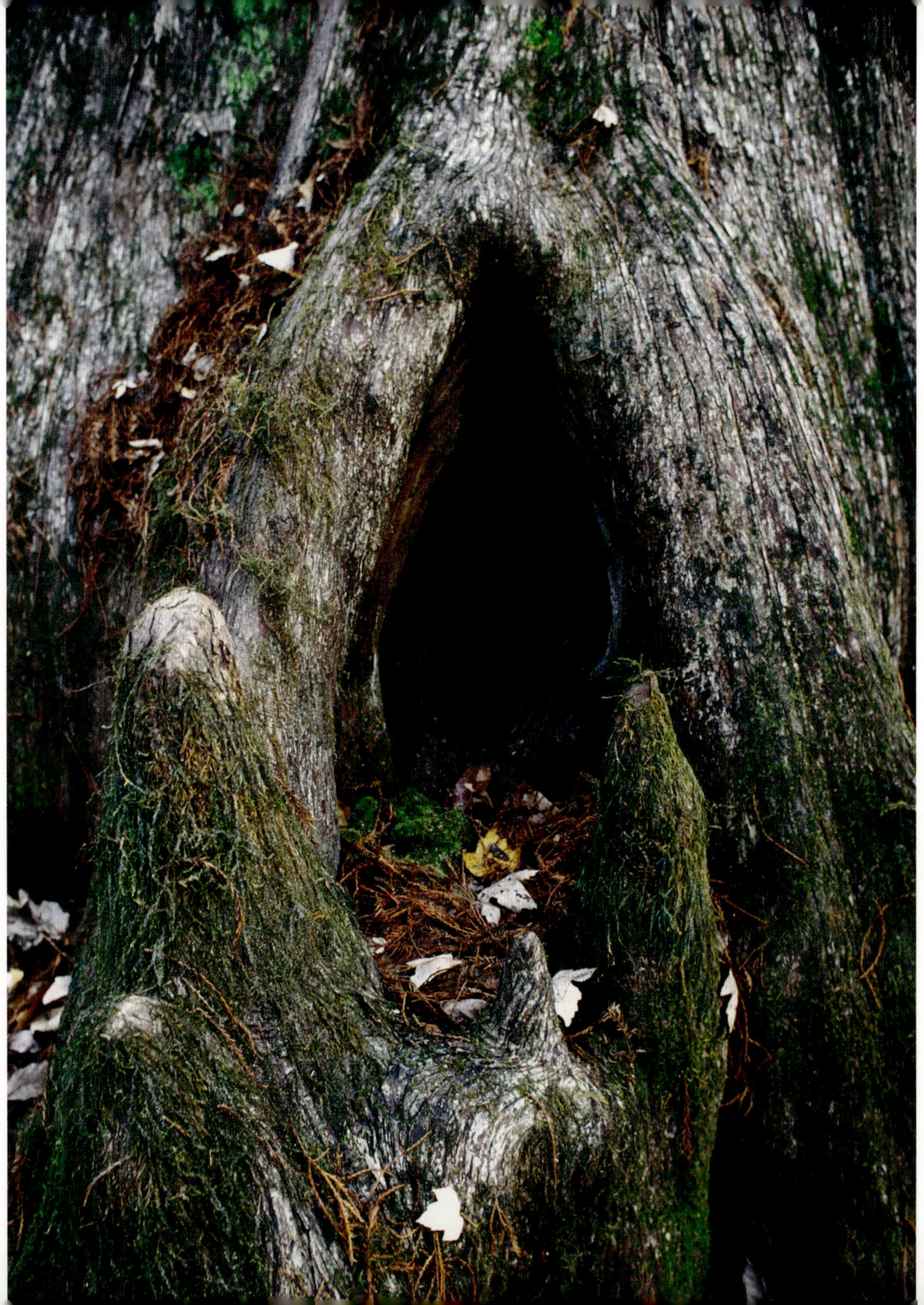

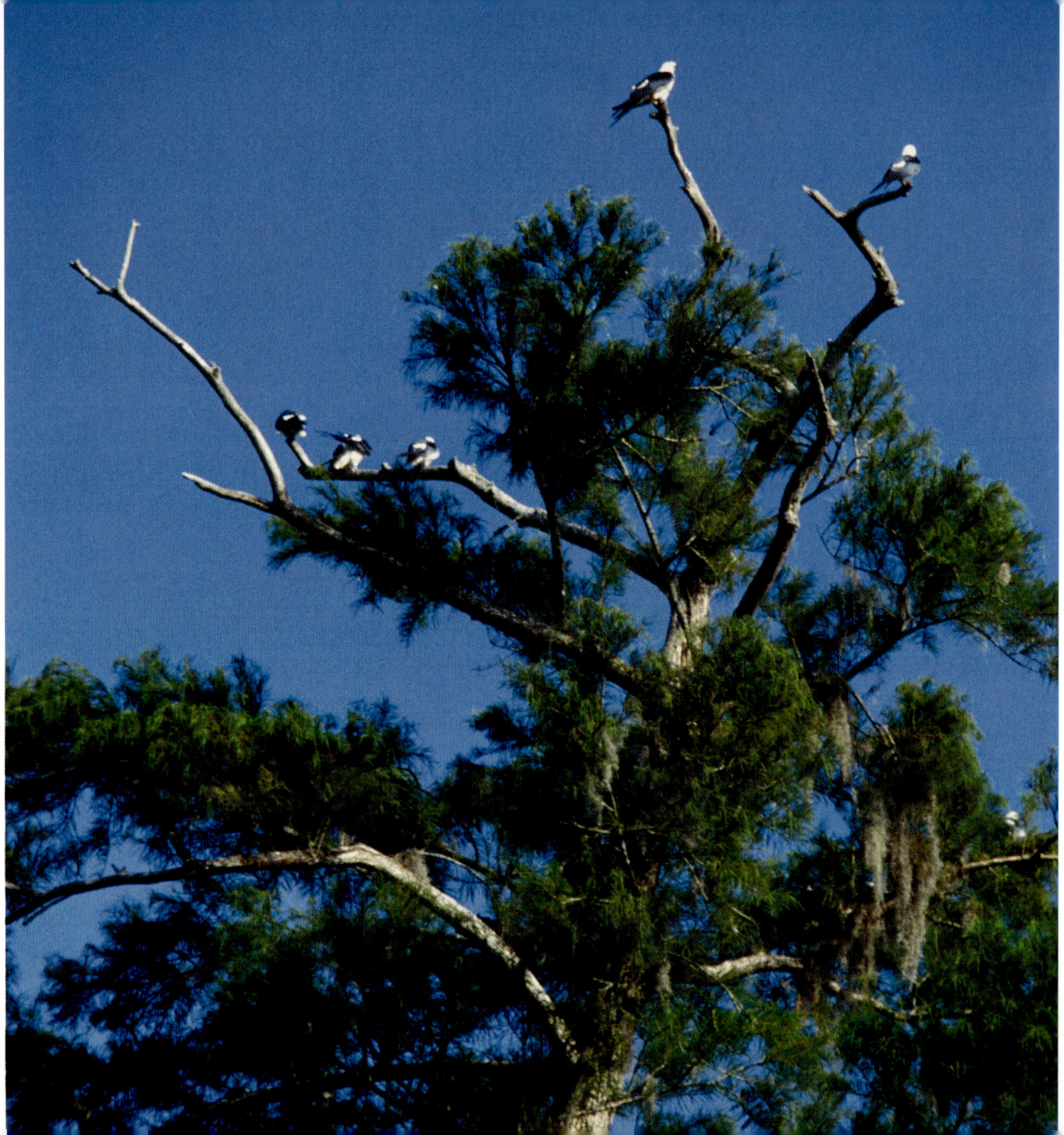

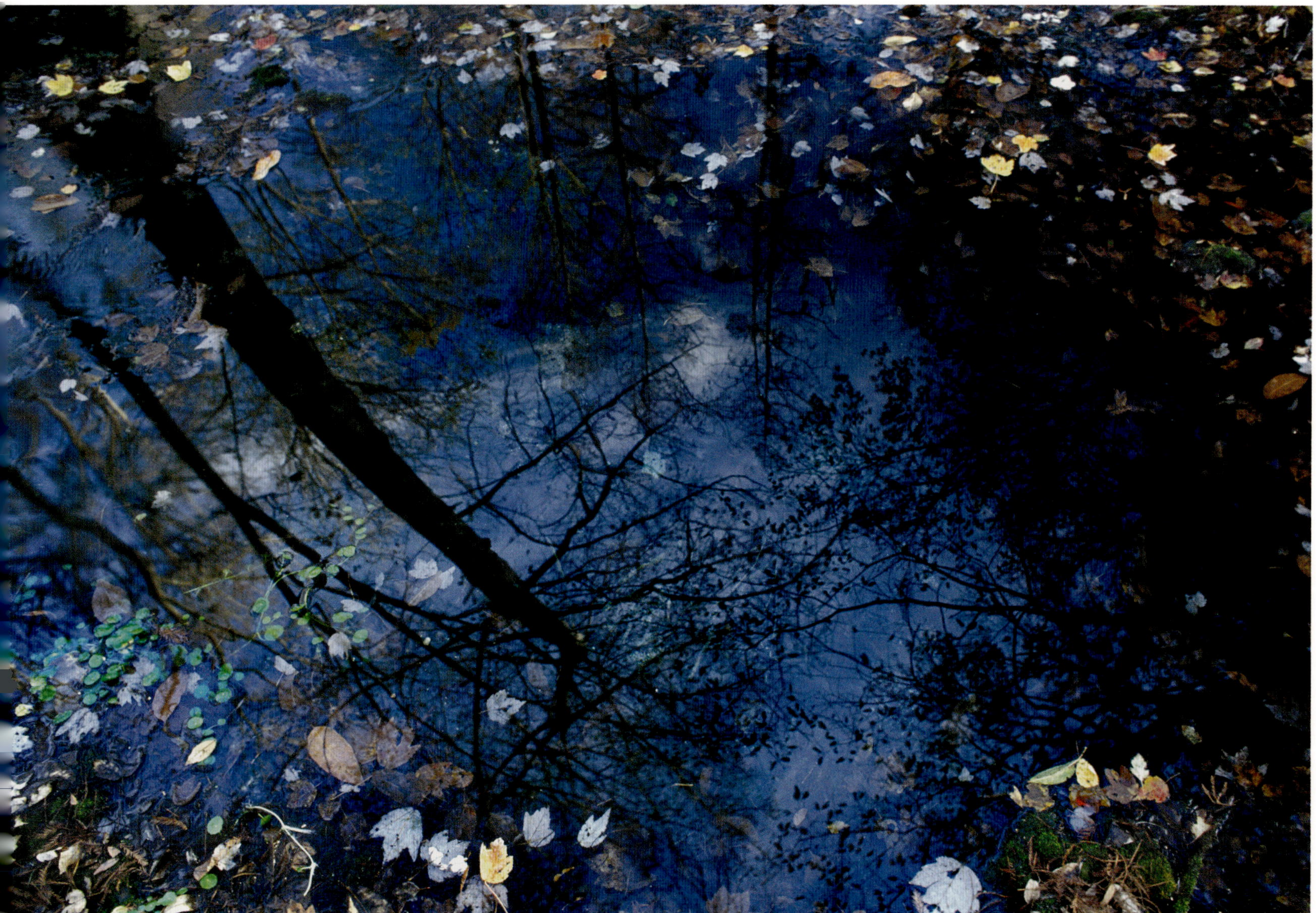

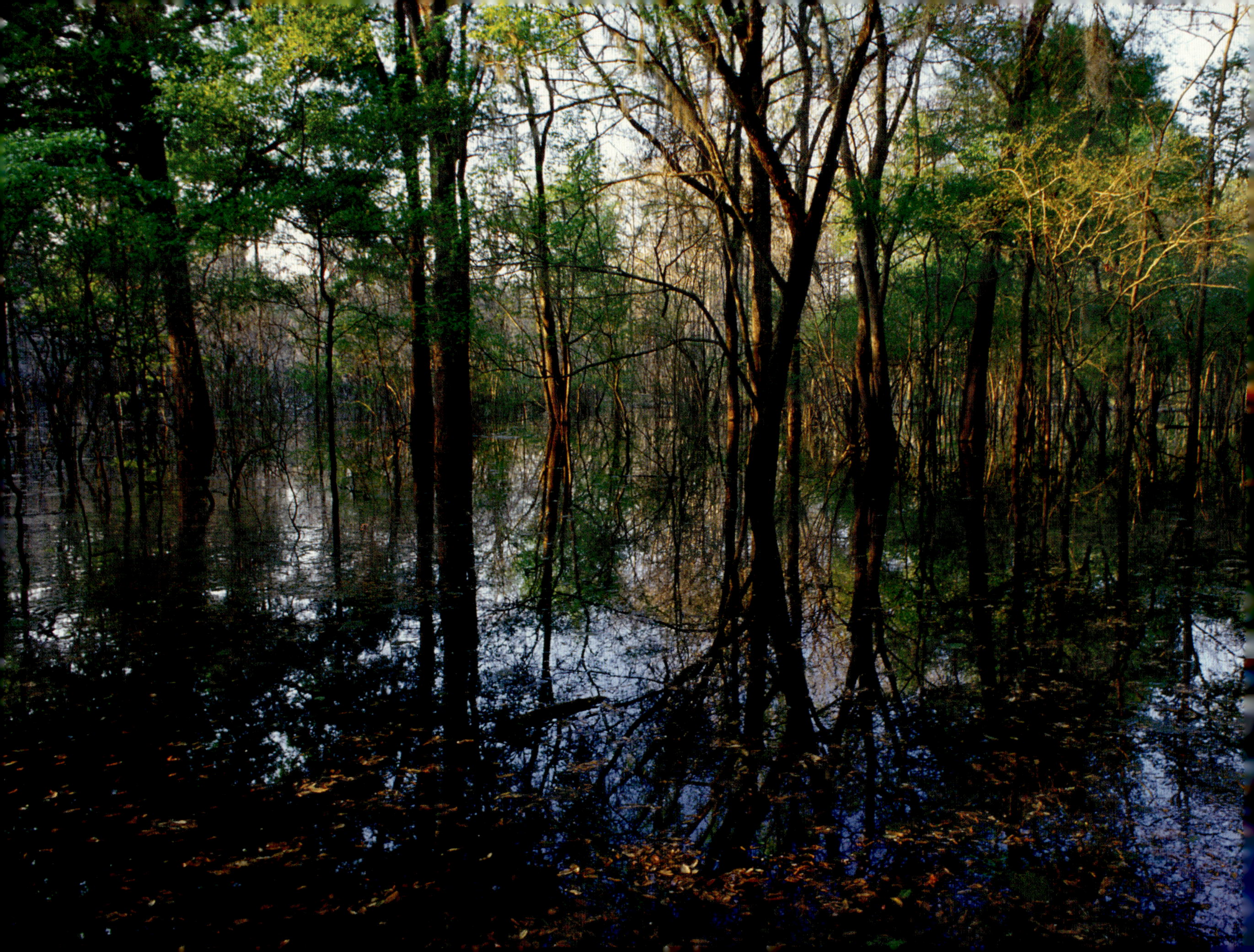

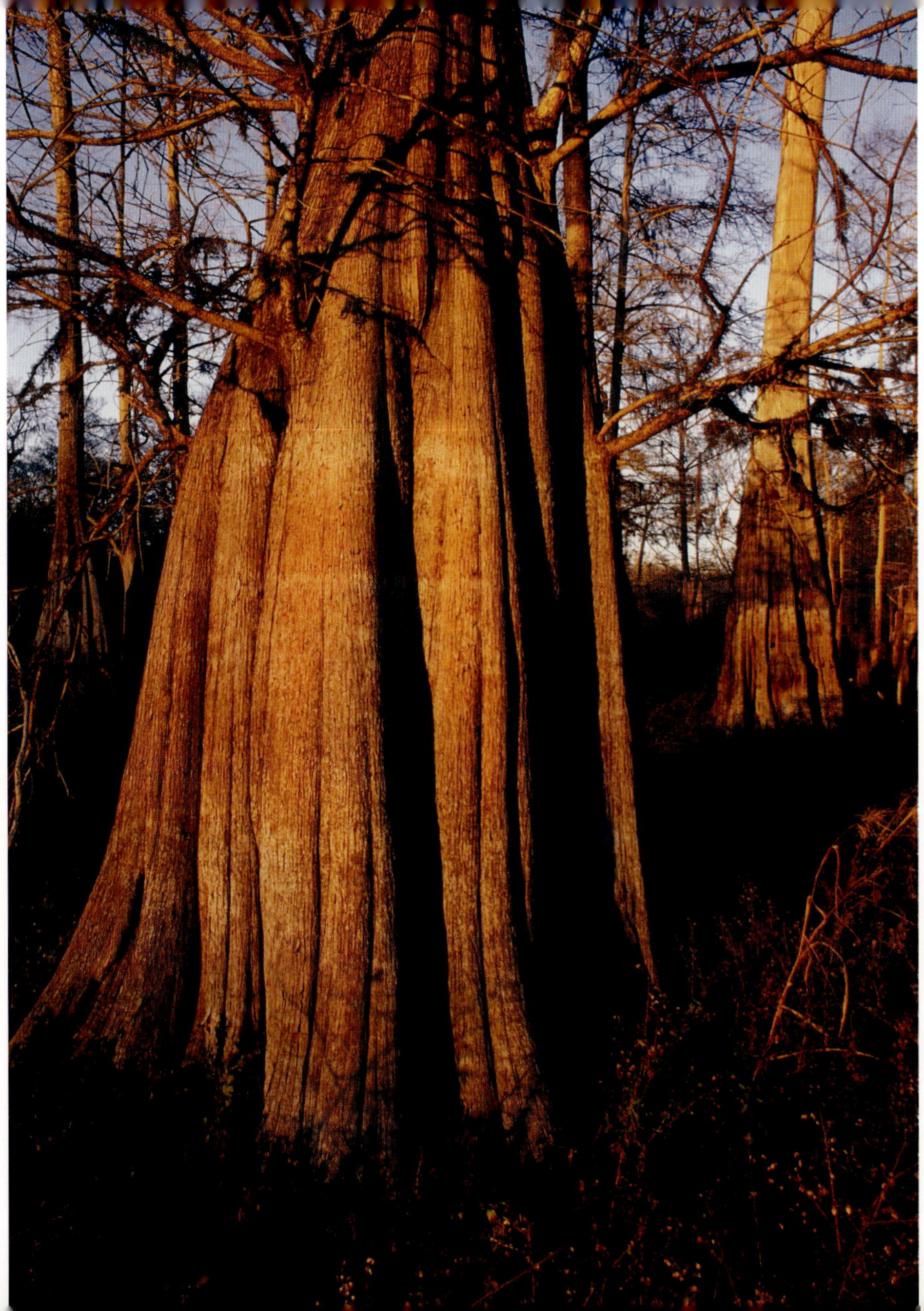

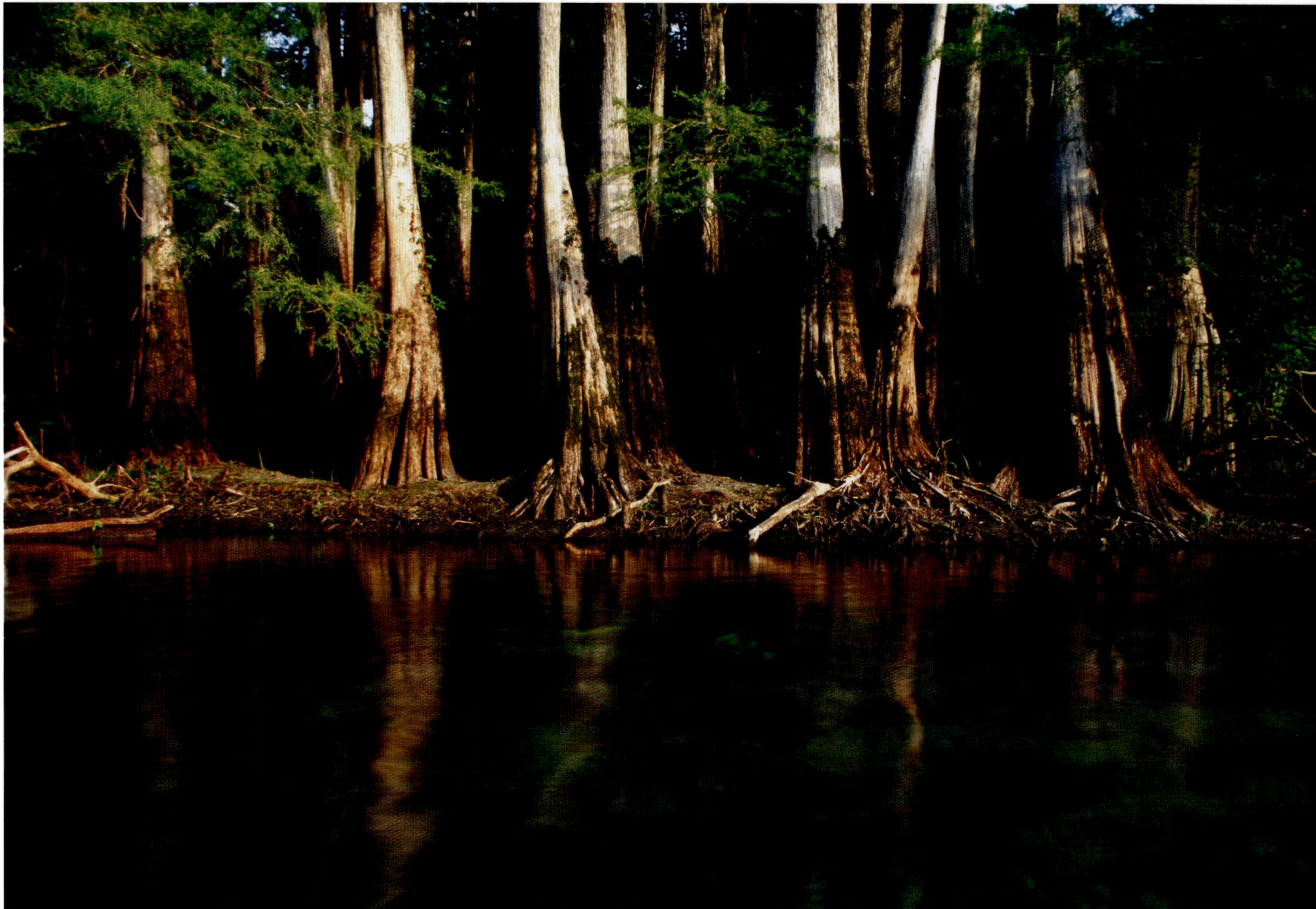

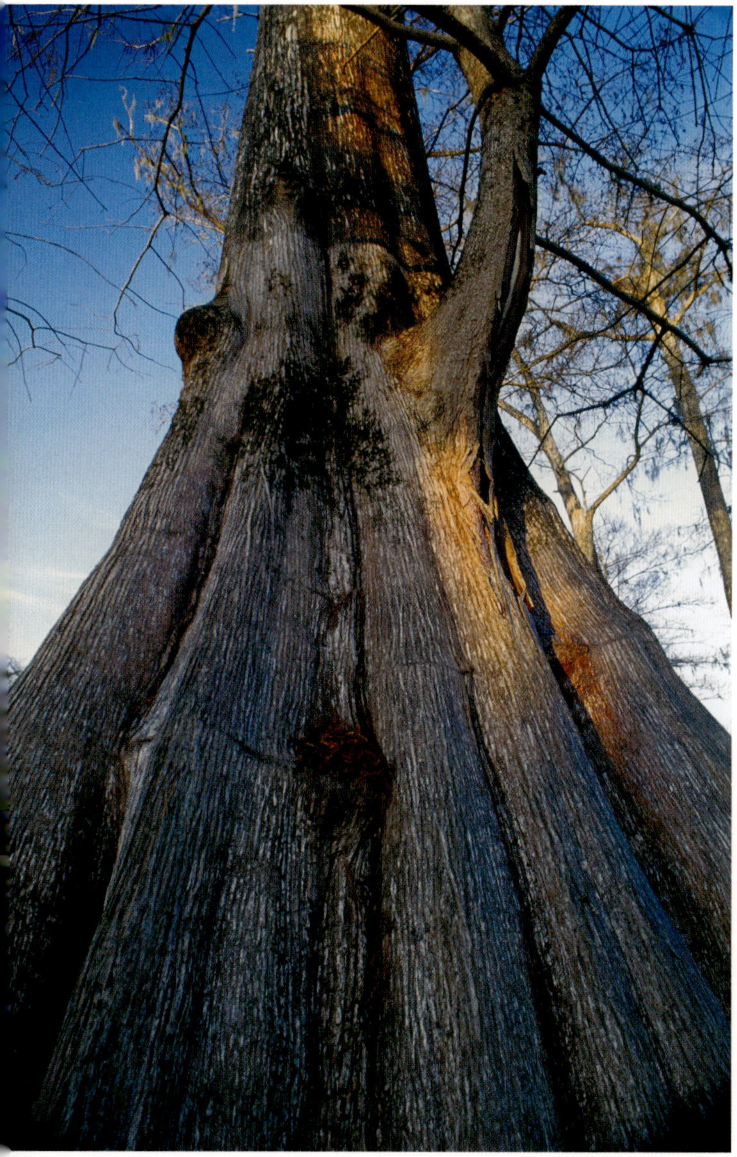

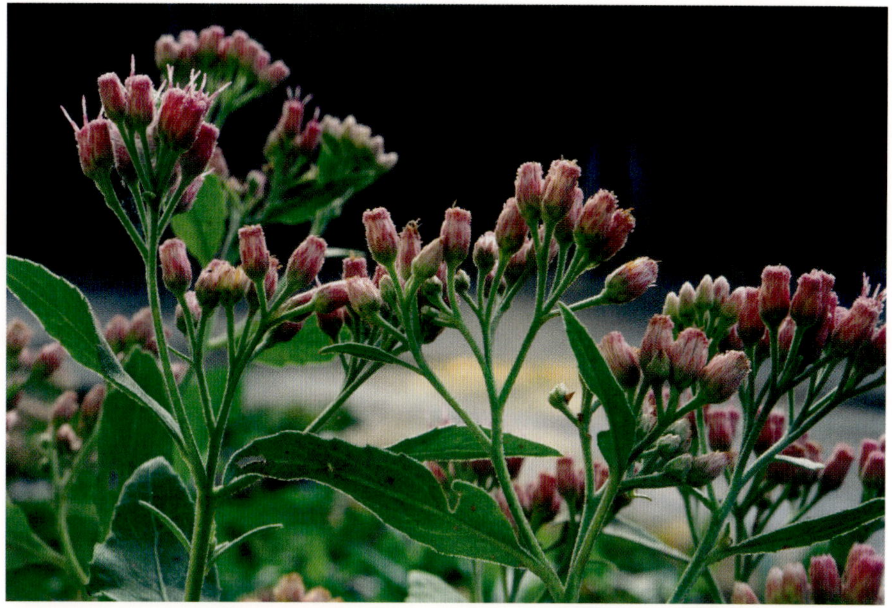

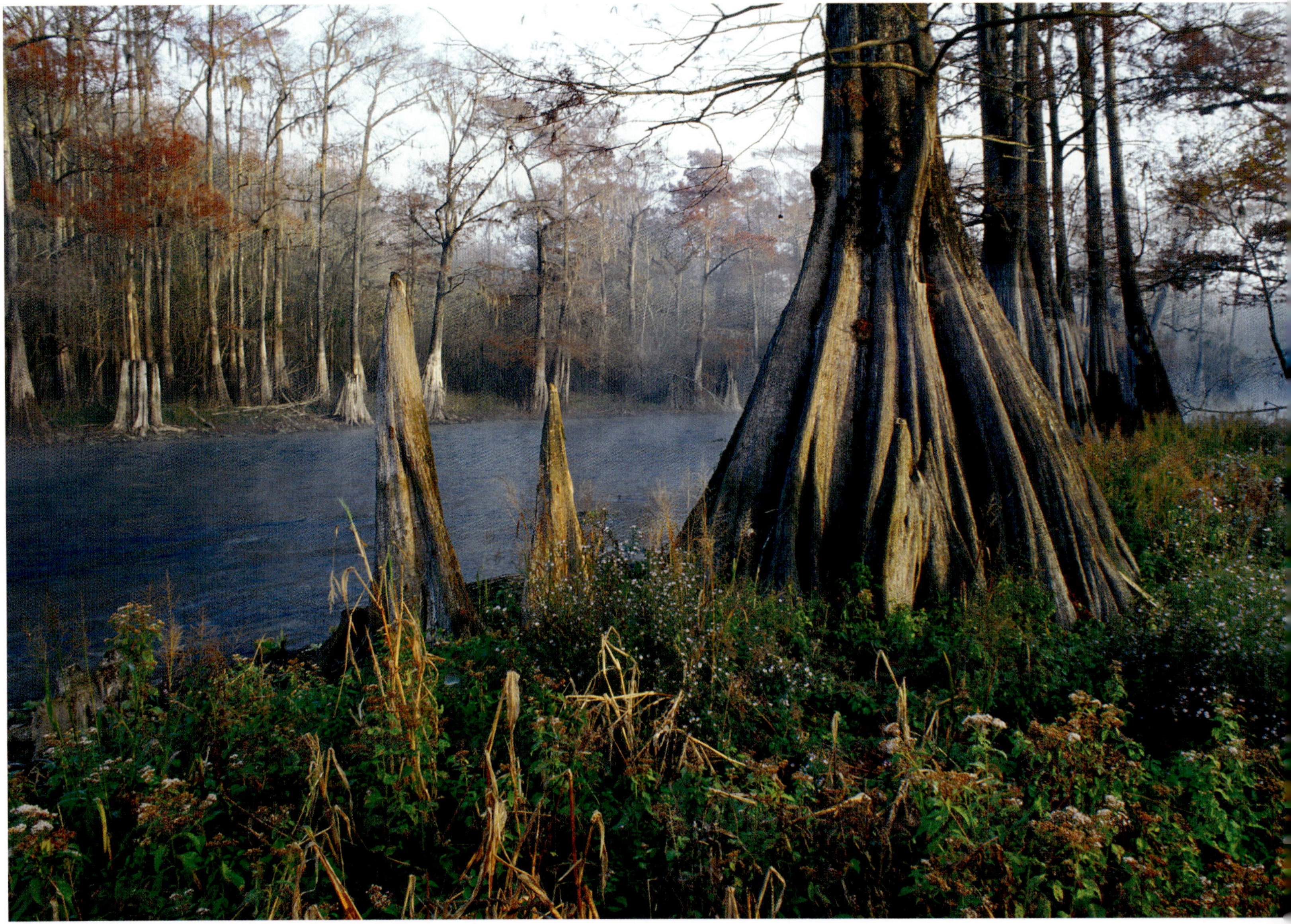

come
walk with me
where no one goes
we will travel silent
as blue mist
seeping into the ancient sleep of cypress
where the river flows

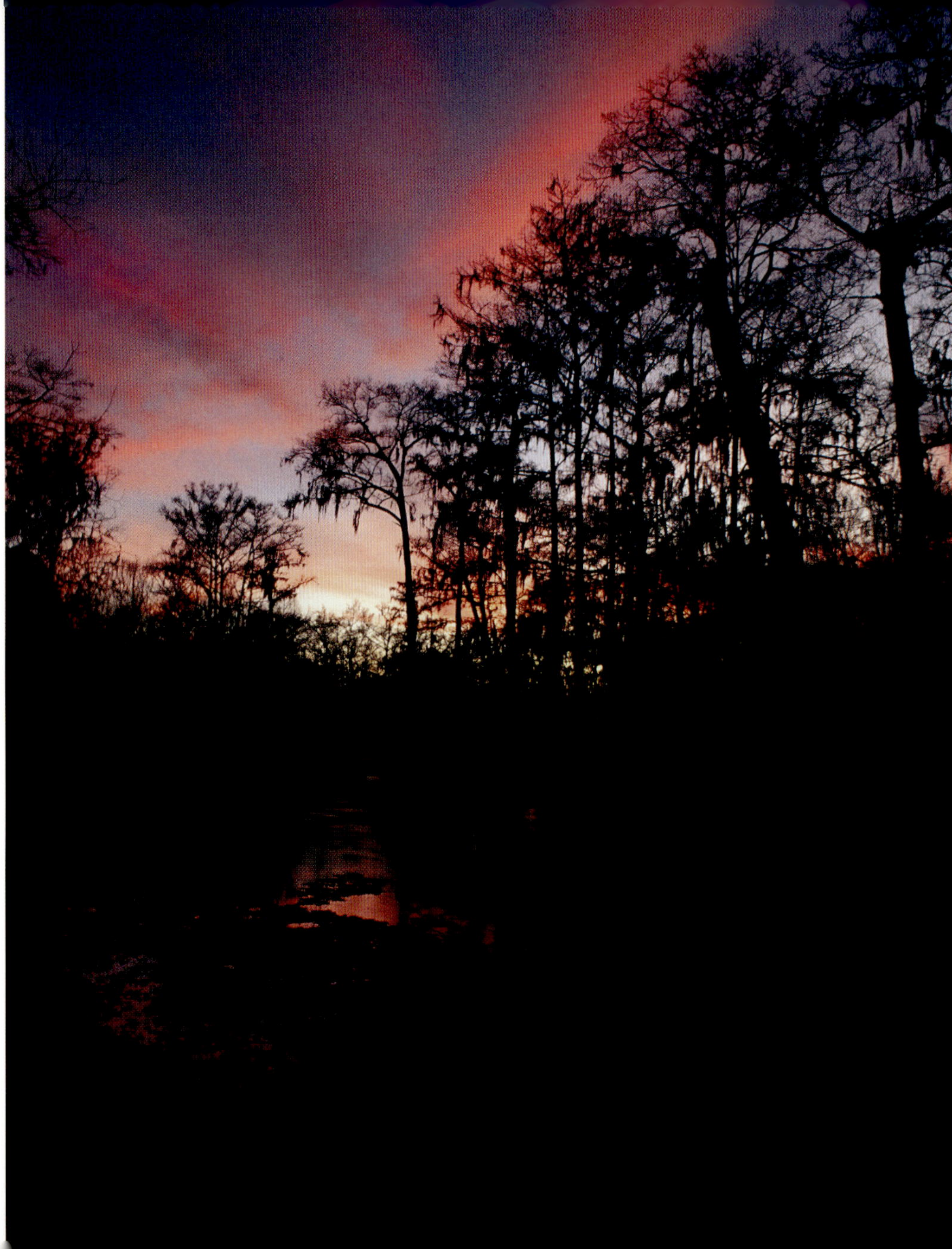

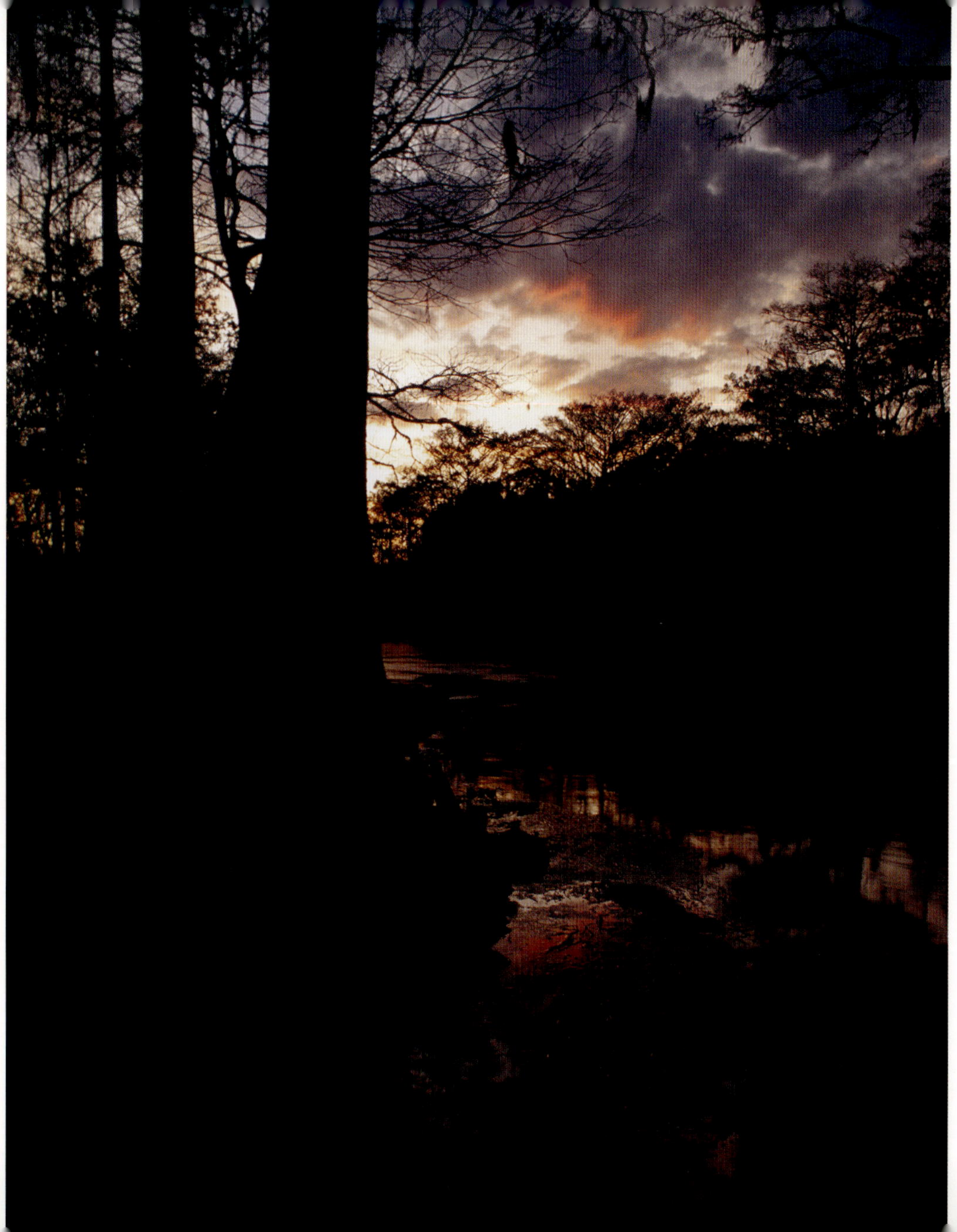

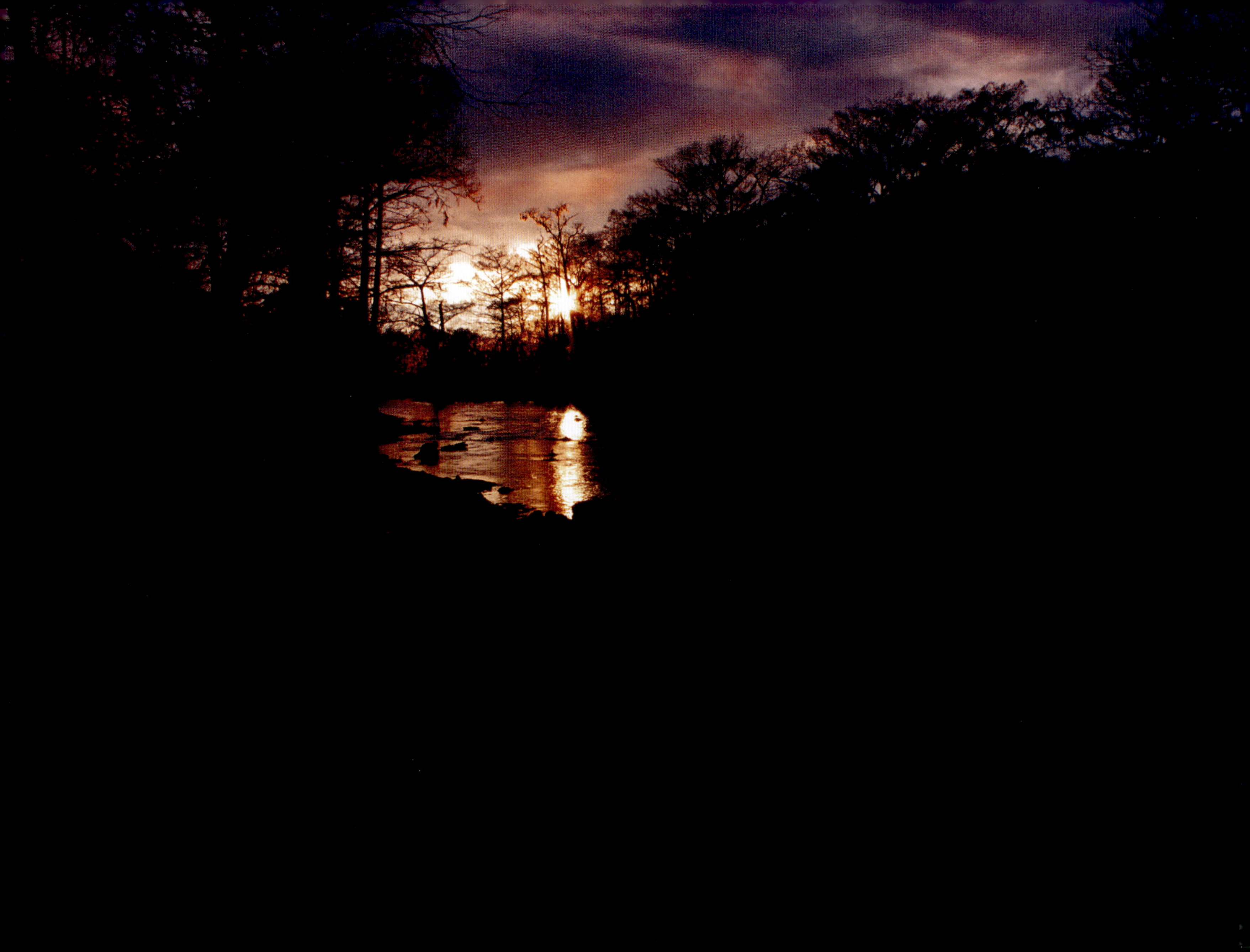

Afterword

Jim Stevenson

Decades ago, a man named Archie Carr traveled throughout Florida and the Caribbean, earning a reputation as an outspoken naturalist by writing about the plight of sea turtles and the wonders of the subtropical world. When Carr saw the Ichetucknee River, he called it "the most beautiful landscape in the world." What might Carr say today, though, if he knew—as we do—that this magnificent treasure is in trouble?

The book you hold is a celebration of the Ichetucknee, but it would be a disservice to readers and a slight to Steve Earl's work if we ignored the facts. The beauty captured here is vulnerable.

The problem is clear: development and agriculture throughout the spring basin are causing pollutants to enter the groundwater. These pollutants, primarily nitrate, are flowing into the springs that feed the Ichetucknee. They are changing the natural chemistry of the Ichetucknee and its profusion of life. A casual visitor to the river might not notice any of this, but those who have visited the river for decades can see that the water is changing from clear blue to cloudy green. Numerous bathers and tubers who visited this "pristine" river in recent years have suffered skin irritations and other health effects.

For centuries, the river and human beings got along just fine. As early as 1539, Hernando de Soto enjoyed a dip in the river when he stayed at an Indian village during his explorations through Florida. Priests at San Martin de Timucua bathed in the river in the 1600s. The 1800s brought limited development along the river, including phosphate mining, a grist mill, a general store, and a post office; but all this was insignificant compared to the impact of housing, commercial development, and agriculture in the 1900s. What scientists now know is that threats to the river come from a three-hundred-square-mile hydrological basin, the underground watershed that feeds the springs and river. The State of Florida protected the river in 1970 when it created the Ichetucknee Springs State Park. As many as two hundred thousand visitors enjoy the park each year, but threats to the springs and river lie beyond the park boundaries.

Rainfall recharges the springs that feed the Ichetucknee. Rainwater falling in developed areas of Columbia County and Lake City takes with it pollutants. Research over the past twelve years indicates that nearly three hundred tons of nitrates are flowing from the land and into the aquifer each year. Specifically, nitrates flow with water out of septic tanks and off of livestock pastures. Rains cause fertilizers to leach out of lawns and off golf courses. The nitrates cause algae to grow at abnormal rates. The algae growth depletes oxygen and chokes the growth of the river's native species, such as eelgrass. This sets off a chain of events that impacts many other aspects of the river's health.

The area of the basin most vulnerable to contamination is that between I-75 and the springs. The Ichetucknee trace, the former riverbed, is especially vulnerable because a major underground cave that feeds the springs lies beneath the trace. Septic tanks in this area, as well as areas surrounding sinkholes connected to caves, are of special concern.

The fix is not cheap. Nitrogen-removing septic systems are more expensive than the standard septic tanks. However, as more homes are constructed in the basin, human wastes must be managed in a better way. As populations increase, society must make adjustments to account for the added pollution we create. There was a time when homeowners were required to abandon their outhouses and install septic tanks to protect drinking water from bacteria. Now it's time to improve septic tanks to control nitrates.

The future of the Ichetucknee lies in the ability of residents and local government to act. The good news is that many are trying. The people of Columbia County are taking action. The point is not to assess blame, but to commend successes. The Florida Springs Task Force, for example, has recognized the good deeds of several Lake City businesspeople, officials, and citizens by presenting them the Florida Springs Protection Award. There are too many successes to mention here, but you can see the impressive list of "2008 Accomplishments" at www.ichetucknee.info.

There are other groups and initiatives at work. Local churches, for example, have been holding "Baptism Sundays" at the river, helping foster a deeper spiritual connection between citizens and the river. Also, the Ichetucknee Springs Old Timers Days have been held at the headspring annually for the past eleven years. Anyone who knew the Ichetucknee before the state park was established in 1970 is an official "Old-Timer." They are interviewed to record their stories of the good old days when the springs were a center of social activity for the community.

The public is welcome to attend the quarterly meetings of the Ichetucknee Springs Basin Working Group held in Lake City. The meeting agenda is listed on the working group Web site: www.ichetucknee.info. Or call the working group coordinator at (850) 562-5716 for information.

Everyone is also invited to Ichetucknee Springs Appreciation Day, where hydrologists, geologists, biologists, paleontologists, and archaeologists set up exhibits to educate the public on all aspects of the river's health and history. Admission is free on this day for all Columbia County residents, for it is only through local initiatives that the springs and river can stay healthy.

Indeed, saving the Ichetucknee is a local matter. Although we can look to Tallahassee and to Washington, D.C., to provide a portion of the needed funding, we can't expect state and federal governments to fix this local problem. The solutions must be developed and implemented first and foremost by the residents of the Ichetucknee basin, with the additional help of all concerned citizens.

Nearly seven million people have visited the Ichetucknee since the state park was established. More than any other Florida spring, the Ichetucknee is the "people's spring." Numerous research projects are in the works to identify and overcome pollution sources. Those who have floated these sparkling waters care deeply. It will not be cheap or easy, but what is the alternative when it comes to saving what Carr recognized, along with so many of us, as one of the world's most beautiful places?

The most powerful legacy of this artful book will be if the world it captures is saved, for generations to come, on more than just paper.

Resources

Ichetucknee Springs Basin Working Group
4797 Lakely Drive, Tallahassee, Florida 32303
floridasprings@comcast.net

Save Our Suwannee, Inc.
P.O. Box 669
Bell, Florida 32619
yenrab@alltel.net

4 Rivers Audubon
P.O. Box 596
Lake City, Florida 32056
jsulek214@earthlink.net

Ichetucknee Springs State Park
12087 SW U.S. Highway 27
Fort White, Florida 32038

I only went out for a walk and finally concluded to stay out till sundown, for going out, I found, was really going in.

John Muir

Acknowledgments

With great respect and love, I wish to thank the following people for their direct contributions to the realization of this book. You are each a special part of this story and it would not have come to light without you all: Loye Barnard, Al Burt, John Byram, Ellen Carroll, Susan Cerulean, Ian Cole, Ellie Corey, Johnny Dame, Tim Davis, Crystal Earl, Emily Earl, Janaka Earl, John Earl Sr., John Earl Jr., Kamala Earl, Larry Earl, Leonora Earl, Saroj Earl, Susan Tinkelman Earl, Karen Entner, Kristin Fletcher, Bill Hanrahan, Rick Hughes, Ichetucknee Springs State Park staff, John Knaub, Mark Kovitch, John Moran, Enoch Nadler, Noah Nadler, Sam Nadler, Cathy Nagler, Jane O'Farroll, Tim Ohr, Annie Pais, Ron Preston, Renée Rau, S. G. C., Mickey Singer, Jim Stevenson, Bahira Sugarman, Craig Swearingen, University Press of Florida staff, Jim Valentine, and Gary Weinkle.

Over the course of my life, many more people have offered their unique gifts of love, support, and inspiration, which have nurtured this body of work. My heartfelt thanks to each and every one of you!

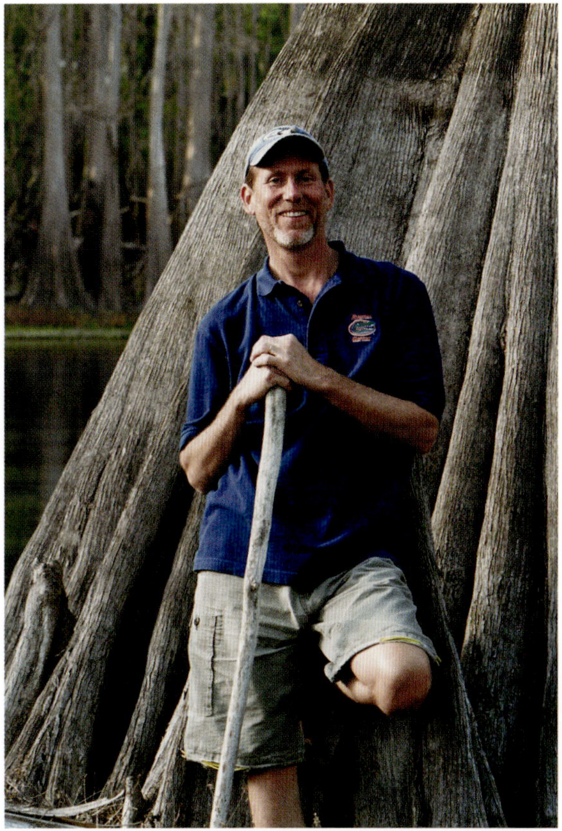

Steven Earl lives and works as a ranger at Gold Head Branch State Park in Keystone Heights, Florida. He heads the park's volunteer, interpretive, and educational outreach programs. He has developed a series of photographic shows that have been presented at school, library, and civic events. His photography has been featured in the *Florida's Fabulous* nature book series and a variety of other publications.